Graduate Texts in Mathematics 245

Graduate Texts in Mathematics

(continued after index)

Jane P. Gilman
Irwin Kra
Rubí E. Rodríguez

Complex Analysis

In the Spirit of Lipman Bers

 Springer

Jane P. Gilman
Department of
    Mathematics and
    Computer Science
Rutgers University
Newark, NJ 07102
USA
gilman@rutgers.edu

Irwin Kra
Math for America
50 Broadway, 23 Floor
New York, NY 10004
and
Department of Mathematics
State University of New York
    at Stony Brook
Stony Brook, NY 11794
USA
irwin@math.sunysb.edu

Rubí E. Rodríguez
Pontificia Universidad
    Católica de Chile
Santiago
Chile
rubi@mat.puc.cl

ISBN: 978-0-387-74714-9   e-ISBN: 978-0-387-74715-6

Library of Congress Control Number: 2007940048

Mathematics Subject Classification (2000): 30-xx 32-xx

Printed on acid-free paper.

9 8 7 6 5 4 3 2 1

springer.com

To the memory of

Mary and Lipman Bers

Preface

This book presents fundamental material that should be part of the education of every practicing mathematician. This material will also be of interest to computer scientists, physicists, and engineers.

Complex analysis is also known as function theory. In this text we address the theory of complex-valued functions of a single complex variable. This is a prerequisite for the study of many current and rapidly developing areas of mathematics, including the theory of several and infinitely many complex variables, the theory of groups, hyperbolic geometry and three-manifolds, and number theory. Complex analysis has connections and applications to many other many other subjects in mathematics, and also to other sciences as an area where the classic and the modern techniques meet and benefit from each other. We will try to illustrate this in the applications we give.

Because function theory has been used by generations of practicing mathematicians working in a number of different fields, the basic results have been developed and redeveloped from a number of different perspectives. We are not wedded to any one viewpoint. Rather we will try to exploit the richness of the subject and explain and interpret standard definitions and results using the most convenient tools from analysis, geometry, and algebra.

The key first step in the theory is to extend the concept of differentiability from real-valued functions of a real variable to complex-valued functions of a complex variable. Although the definition of complex differentiability resembles the definition of real differentiability, its consequences are profoundly different. A complex-valued function of a complex variable that is differentiable is called *holomorphic* or *analytic*, and the first part of this book is a study of the many equivalent ways of understanding the concept of analyticity. Many of the equivalent ways of formulating the concept of an analytic function are summarized in what we term the Fundamental Theorem for functions of a complex variable. Chapter 1 begins with two motivating examples, followed by the statement of the Fundamental Theorem, an outline of the plan for proving it, and a description of the text contents: the plan for the rest of the book.

In devoting the first part of this book to the precise goal of stating and proving the Fundamental Theorem, we follow a path charted for us by Lipman Bers, from whom we learned the subject. In his teaching, expository, and research writing he often started by introducing a main, often technical, result and then proceeded to derive its important and seemingly surprising consequences. Some of the grace and elegance of this subject will not emerge until a more technical framework has been established. In the second part of the text, we proceed to the leisurely exploration of interesting ramifications and applications of the Fundamental Theorem.

We are grateful to Lipman Bers for introducing us to the beauty of the subject. The book is an outgrowth of notes from Bers's original lectures. Versions of these notes have been used by us at our respective home institutions, some for more than 20 years, as well as by others at various universities. We are grateful to many colleagues and students who read and commented on these notes. Our interaction with them helped shape this book. We tried to follow all useful advice and correct, of course, any mistakes or shortcomings they identified. Those that remain are entirely our responsibility.

Jane Gilman
Irwin Kra
Rubí E. Rodríguez

June 2007

Standard Notation and Commonly Used Symbols

A LIST OF SYMBOLS

TERM	MEANING
$\mathbb{Z}$	integers
$\mathbb{Q}$	rationals
$\mathbb{R}$	reals
$\mathbb{C}$	complex numbers
$\widehat{\mathbb{C}}$	$\mathbb{C} \cup \{\infty\}$
$\imath$	a square root of -1
$\imath\mathbb{R}$	the imaginary axis in $\mathbb{C}$
$\Re z$	real part of z
$\Im z$	imaginary part of z
$z = x + \imath y$	$x = \Re z$ and $y = \Im z$
$\bar{z}$	conjugate of z
$r = |z|$	absolute value of z
$\theta = \arg z$	an argument of z
$z = re^{\imath\theta}$	$r = |z|$ and $\theta = \arg z$
$u_x,\ u_y$	real partial derivatives
$u_z,\ u_{\bar{z}}$	complex partial derivatives
$\frac{\partial f}{\partial x}$	partial derivative
∂R	boundary of set R
$|R|$	cardinality of set R
$\operatorname{cl} R$	closure of set R
$\operatorname{int} R$	interior of set R
$X_{condition}$	the set of $x \in X$ that satisfy *condition*
$\nu_\zeta(f)$	order of the function f at the point ζ
$i(\gamma)$	interior of the Jordan curve γ
$e(\gamma)$	exterior of the Jordan curve γ
$f|_B$	restriction of the function f to the subset B of its domain
$U(z,r) = U_z(r)$	$\{\zeta \in \mathbb{C};\ |\zeta - z| < r\}$
$\mathbb{D}$	$U(0,1)$
$\mathbb{H}^2$	$\{z \in \mathbb{C};\ \Im z > 0\}$

STANDARD TERMINOLOGY

TERM	MEANING
LHS	left-hand side
RHS	right-hand side
deleted neighborhood of z	neighborhood with z removed
CR	Cauchy Riemann equations
$\subset$	proper subset
$\subseteq$	subset, may not be proper
d	Euclidean distance on $\mathbb{C}$
ρ_D	hyperbolic distance on D
MMP	Maximum Modulus Property
MVP	Mean Value Property

Contents

CHAPTER 1

The Fundamental Theorem in Complex Function Theory

This introductory chapter is meant to convey the need for and the intrinsic beauty found in passing from a real variable x to a complex variable z. In the first section we "solve" two natural problems using complex analysis. In the second, we state the most important result in the theory of functions of one complex variable that we call the Fundamental Theorem of complex variables; its proof will occupy most of this volume. The next-to-last section of this chapter is an outline of our plan for the proof; in subsequent chapters, we will define all the concepts encountered in the statement of the theorem. Therefore, the reader may not be able at this point to understand all (or any) of the statements in the theorem, to fully appreciate the two motivating examples, or to appreciate the depth of the various claims of the theorem and might choose initially to skim this material. All readers should periodically, throughout the journey through this book, return to this chapter. Finally we end this chapter with a section that gives a more conventional outline of the text.

1.1. Some motivation

1.1.1. Where do series converge? In the calculus of a real variable one encounters two series that converge for $|x| < 1$ but in no larger open interval:

$$\frac{1}{1+x} = 1 - x + x^2 - \ldots + (-1)^n x^n + \ldots$$

and

$$\frac{1}{1+x^2} = 1 - x^2 + x^4 - \ldots + (-1)^n x^{2n} + \ldots .$$

It is natural to ask why these two series that are centered at the origin have radius of convergence 1. The answer for the first one is natural: the function $\frac{1}{1+x}$ has a singularity at $x = -1$, and so the series certainly cannot represent the function at this point, which is at distance 1 from 0. For the second series, the answer does not appear readily within real analysis. However, if we view $\frac{1}{1+x^2}$ as a function of the complex

1

variable[1] x, then we again conclude that the series representing this function should have radius of convergence 1, since that is the distance from 0 to the singularities of the function; they are at $\pm i$.

1.1.2. A problem on partitions. A natural question in elementary additive number theory is the following: Is it possible to partition the positive integers $\mathbb{Z}_{>0}$ into finitely many (more than 1) infinite arithmetic progressions with distinct differences? The answer is "NO". The assumption on distinct differences is clearly necessary. So assume to the contrary that

$$\mathbb{Z}_{>0} = S_1 \cup S_2 \cup \ldots \cup S_n,$$

where $n \in \mathbb{Z}_{>1}$, and for $1 \le i \le n$, S_i is an arithmetic progression with initial term a_i and difference d_i, for $1 \le i < j \le n$, $S_i \cap S_j = \emptyset$, and $1 < d_1 < d_2 < \ldots < d_n$. Then

$$\sum_{i=1}^{\infty} z^i = \sum_{i\in S_1} z^i + \sum_{i\in S_2} z^i + \ldots + \sum_{i\in S_n} z^i,$$

and each series converges for $|z| < 1$. Summing the above geometric series, we see that

$$\frac{z}{1-z} = \frac{z^{a_1}}{1-z^{d_1}} + \frac{z^{a_2}}{1-z^{d_2}} + \ldots + \frac{z^{a_n}}{1-z^{d_n}} \text{ for all } z \text{ with } |z| < 1. \quad (1.1)$$

Choose a sequence of complex numbers[2] $\{z_k\}$ of absolute value less than 1 with $\lim_{k\to\infty} z_k = e^{\frac{2\pi i}{d_n}}$. Then

$$\lim_{k\to\infty} \frac{z_k}{1-z_k} = \frac{e^{\frac{2\pi i}{d_n}}}{1-e^{\frac{2\pi i}{d_n}}}$$

and

$$\lim_{k\to\infty} \frac{z_k^{a_i}}{1-z_k^{d_i}} = \frac{e^{\frac{2\pi i a_i}{d_n}}}{1-e^{\frac{2\pi i d_i}{d_n}}} \text{ for } i = 1, 2, \ldots, n-1;$$

(all these quantities are finite), whereas $\lim_{k\to\infty} \frac{z_k^{a_n}}{1-z_k^{d_n}}$ does not exist. This is an obvious contradiction to (1.1).

---

[1] In the sequel we usually use z, w, and ζ, among others, but not x to denote a complex variable.

[2] Notation for the polar form of a complex number is established in Chapter 2.

1.2. The Fundamental Theorem

THEOREM 1.1. *Let $D \subseteq \mathbb{C}$ denote a domain (an open connected set), and let $f = u + iv : D \to \mathbb{C}$ be a complex-valued function defined on D.*

The following conditions are equivalent:

(1) The complex derivative

$$f'(z) \ \text{exists for all } z \in D; \qquad \text{(Riemann)}$$

that is, the function f is holomorphic on D.

(2) The functions u and v are continuously differentiable and satisfy

$$\frac{\partial u}{\partial x} = \frac{\partial v}{\partial y} \ \text{and} \ \frac{\partial u}{\partial y} = -\frac{\partial v}{\partial x}. \qquad \text{(Cauchy–Riemann: CR)}$$

Alternatively, the function f is continuously differentiable and satisfies

$$\frac{\partial f}{\partial \overline{z}} = 0. \qquad \text{(CR–complex form)}$$

(3) For each simply connected subdomain $\widetilde{D}$ of D there exists a holomorphic function $F : \widetilde{D} \to \mathbb{C}$ such that $F'(z) = f(z)$ for all $z \in \widetilde{D}$.

(4) The function f is continuous on D, and if γ is a (piecewise smooth) closed curve in a simply connected subdomain of D, then

$$\int_{\gamma} f(z)dz = 0.$$

((1) $\Rightarrow$ (4): Cauchy's theorem; (4) $\Rightarrow$ (1): Morera's theorem)
An equivalent formulation of this condition is as follows: The function f is continuous, and the differential form $f(z)dz$ is closed on D.

(5) If $\{z \in \mathbb{C} : |z - z_0| \le r\} \subseteq D$, then

$$f(z) = \frac{1}{2\pi i} \int_{|\tau - z_0| = r} \frac{f(\tau)}{\tau - z} \, d\tau \qquad \text{(Cauchy's Integral Formula)}$$

for each z such that $|z - z_0| < r$.

(6) The n-th complex derivative

$$f^{(n)}(z) \ \text{exists for all } z \in D \ \text{and for all integers } n \ge 0.$$

(7) *If* $\{z : |z - z_0| \leq r\} \subseteq D$ *with* $r > 0$, *then there exists a unique sequence of complex numbers* $\{a_n\}_{n=0}^{\infty}$ *such that*

$$f(\mathbf{z}) = \sum_{n=0}^{\infty} a_n (\mathbf{z} - z_0)^n \qquad \text{(Weierstrass)}$$

for each $\mathbf{z}$ *such that* $|\mathbf{z} - z_0| < r$.

Furthermore, the series converges uniformly and absolutely on every compact subset of $\{z : |z - z_0| < r\}$. *The* a_n *may be computed as*

$$a_n = \frac{1}{2\pi i} \int_{|\tau - z_0| = r} \frac{f(\tau)}{(\tau - z_0)^{n+1}} \, d\tau \qquad \text{(Cauchy)}$$

and

$$a_n = \frac{f^{(n)}(z_0)}{n!}. \qquad \text{(Taylor)}$$

(8) *Choose a point* $z_i \in K_i$, *where* $\bigcup_{i \in I} K_i$ *is the connected component decomposition of the complement of* D *in* $\mathbb{C} \cup \{\infty\}$. *Let* $S = \{z_i; i \in I\}$. *Then the function* f *is the limit (uniform on compact subsets of* D) *of a sequence of rational functions with singularities only in* S.

(Runge's theorem)

1.3. The plan for the proof

We prove the Fundamental Theorem by showing the following implications.

$$(1) \Leftrightarrow (2) \Rightarrow (3) \Rightarrow (4) \Rightarrow (5) \Rightarrow (6) \Rightarrow (1);$$

$$(5) \Rightarrow (7) \Rightarrow (1) \Leftrightarrow (8).$$

It is of course possible to follow other paths through the various claims to obtain our main result. For the convenience of the reader, we describe where the various implications are to be found. At times the reader will need to slightly enhance an argument to obtain the required implication.

(1) $\Leftarrow$ (2): Corollary 2.36.
(1) $\Rightarrow$ (2): Theorem 2.32 and Corollary 5.7.
(1) $\Rightarrow$ (3): Theorem 4.52 and Corollary 4.44.
(3) $\Rightarrow$ (4): This is a trivial implication. See Lemma 4.12 and the definitions preceding it.

$(4) \Rightarrow (5)$: Theorems 5.11 and 5.2.
$(5) \Rightarrow (6)$: The proof of Theorem 5.5.
$(6) \Rightarrow (1)$: This is a trivial implication.
$(5) \Rightarrow (7)$: The proof of Theorem 5.5.
$(7) \Rightarrow (1)$: Theorem 3.17.
$(1) \Rightarrow (8)$: Theorem 7.35.
$(8) \Rightarrow (1)$: Theorem 7.2.

In standard texts, typically each of these implications is stated as a single theorem. The tag words in parentheses are the names or terms that identify the theorems. The forward implication $(1) \Rightarrow (n)$ would be the theorem: "If f is a holomorphic function, then condition (n) holds," where $n \in \{2, 3, 4, 5, 6, 7, 8\}$. For example, $(1) \Rightarrow (2)$ would be stated as, "If f is holomorphic, then the Cauchy–Riemann equations hold." The organization of these conditions (potentially 56 theorems—some trivial) into a single unifying theorem is the hallmark of Bers's mathematical style: clarity and elegance. Here it provides a conceptual framework for results that are highly technical and often computational. The framework comes from insight that, once articulated, will drive the subsequent mathematics and lead to new results.

1.4. Outline of text

Chapter 2 contains the basic definitions. It is followed by a study of power series in Chapter 3. Chapter 4 contains the central material, the Cauchy theory, of the subject. We prove that the class of analytic functions is precisely the same as the class of functions having power series expansions, and we establish other parts of the Fundamental Theorem. Many consequences of the Cauchy theory are established in the next two chapters.

In the second part of the text we proceed to the leisurely exploration of interesting ramifications and applications of the Fundamental Theorem. It starts with an exploration of sequences and series of holomorphic functions in Chapter 7. The Riemann Mapping Theorem and the connection between function theory and hyperbolic geometry are the highlights of Chapter 8. The last two chapters deal with harmonic functions and zeros of holomorphic functions. The latter is the beginning of the deep connections to classic and modern number theory.

CHAPTER 2

Foundations

The first section of this chapter introduces the complex plane, fixes notation, and discusses some useful concepts from real analysis. Some readers may initially choose to skim this section. The second section contains the definition and elementary properties of the class of holomorphic functions - the basic object of our study.

2.1. Introduction and preliminaries

This section is a summary of basic notation, a description of some of the basic properties of the complex number system, and a disjoint collection of needed facts from real analysis (advanced calculus). We remind the reader of some of the formalities behind the standard notation, which we usually approach informally.

We start with some **Notation**: $\mathbb{Z}_{>0} \subset \mathbb{Z} \subset \mathbb{Q} \subset \mathbb{R} \subset \mathbb{C} \subset \widehat{\mathbb{C}}$. Here $\mathbb{Z}$ represents the integers, $\mathbb{Z}_{>0}$ the positive integers,[1] $\mathbb{Q}$ the rationals (the integer n is included in the rationals as the equivalence class of $\frac{n}{1}$), and $\mathbb{R}$ the reals. Whether one views the reals as the completion of the rationals or identifies them with Dedekind cuts, the most important property from the perspective of complex variables is the least upper bound property: Every nonempty set of real numbers that has an upper bound has a least upper bound.

The inclusion of $\mathbb{R}$ into the complex numbers $\mathbb{C}$ needs a bit more explanation. It is specified as follows: For $z \in \mathbb{C}$, we write $z = x + \imath y$ with x and y in $\mathbb{R}$ where the symbol $\imath$ represents a square root of -1 so that $\imath^2 = -1$. With these conventions we can define addition and multiplication of complex numbers[2] using the usual rules for these operation on the reals: For all $x, y, \xi, \eta \in \mathbb{R}$,

$$(x + \imath y) + (\xi + \imath \eta) = (x + \xi) + \imath(y + \eta)$$

and

$$(x + \imath y)(\xi + \imath \eta) = (x\xi - y\eta) + \imath(x\eta + y\xi).$$

---

[1]In general $X_{\text{condition}}$ and $\{x \in X : \text{condition}\}$ will describe the set $x \in X$ that satisfy the condition indicated.

[2]With these operations $(\mathbb{C}, +, \cdot)$ is a field.

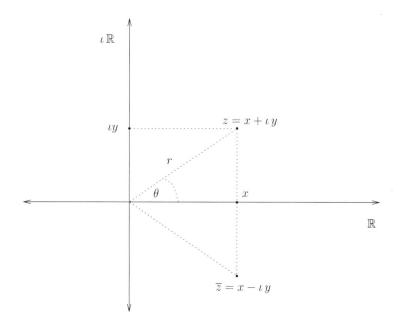

FIGURE 2.1. The complex plane

The *reals* $\mathbb{R}$ are identified with the subset of $\mathbb{C}$ consisting of those numbers with $y = 0$; the *imaginary* numbers $\imath\mathbb{R}$ are those with $x = 0$. For $z = x + \imath y$ in $\mathbb{C}$ with x and y in $\mathbb{R}$, we write $x = \Re z$, the *real part* of z, and $y = \Im z$, the *imaginary part* of z. Geometrically, $\mathbb{R}$ and $\imath\mathbb{R}$ represent the *real* and *imaginary axes* of $\mathbb{C}$, viewed as the complex plane and identified with the cartesian product $\mathbb{R}^2$ or, equivalently, $\mathbb{R} \times \mathbb{R}$.

The complex plane can be viewed as a subset of the complex sphere $\widehat{\mathbb{C}}$, which is $\mathbb{C}$ compactified by adjoining a point, known as *the point at infinity*, so that $\widehat{\mathbb{C}} = \mathbb{C} \cup \{\infty\}$. $\widehat{\mathbb{C}}$ is also called *the extended complex plane* or the *Riemann sphere*. See Exercise 3.18 for a justification of the name.

The complex number $\overline{z} = x - \imath y$ is the *complex conjugate* of the complex number $z = x + \imath y$. Note that $\Re z = \dfrac{z + \overline{z}}{2}$ and $\Im z = \dfrac{z - \overline{z}}{2\imath}$. It is easy to verify the following properties:

Properties of conjugation. For z and $w \in \mathbb{C}$,

(a) $\overline{z + w} = \overline{z} + \overline{w}$,
(b) $\overline{zw} = \overline{z}\,\overline{w}$, and
(c) $\overline{\overline{z}} = z$.

There is a simple and useful **geometric interpretation of conjugation**: It is represented by mirror reflection in real axis. Since $\bar{\bar{z}} = z$, the map $z \mapsto \bar{z}$ defines an involution of $\mathbb{C}$

$$^- : \mathbb{C} \to \mathbb{C} .$$

Another important map, $z \mapsto |z|$ or

$$| \, | : \mathbb{C} \to \mathbb{R}_{\geq 0}$$

is defined by $r = |z| = (z\bar{z})^{\frac{1}{2}} = (x^2 + y^2)^{\frac{1}{2}}$. Here $z = x + \imath y$ and we use the usual convention that unless otherwise specified the square root of a nonnegative number is chosen to be nonnegative. The nonnegative real number r is called the *absolute value* or *norm* or *modulus* of the complex number z.

Properties of absolute value. For z and $w \in \mathbb{C}$,
  (a) $|z| \geq 0$, and $|z| = 0$ if and only if $z = 0$.
  (b) $|zw| = |z| \, |w|$.
  (c) $|z + w| \leq |z| + |w|$. Equality holds whenever either z or w is equal to 0. If $z \neq 0$ and $w \neq 0$, then equality holds if and only if $w = az$ with $a \in \mathbb{R}_{>0}$.
  (d) $|z| = |\bar{z}|$.

Linear representation of $\mathbb{C}$. As a vector space over $\mathbb{R}$, we can identify $\mathbb{C}$ with $\mathbb{R}^2$. Vector addition agrees with complex addition. Scalar multiplication $\mathbb{R} \times \mathbb{C} \to \mathbb{C}$ is the restriction of complex multiplication $\mathbb{C} \times \mathbb{C} \to \mathbb{C}$.

Polar coordinates. A nonzero vector can be described by polar coordinates (r, θ) as well as by the rectangular coordinates (x, y) we have been using. If $z \in \mathbb{C}$ and $z \neq 0$, then we can write

$$z = x + \imath y = r \left(\cos \theta + \imath \sin \theta \right),$$

where $r = |z|$ and $\theta = \arg z$ (an *argument* of z) $= \arcsin \dfrac{y}{r} = \arccos \dfrac{x}{r}$.

Note that the last two identities are needed to define the argument and that[3] $\arg z$ is defined up to the addition of an integral multiple of 2π.

If $w = \rho[\cos \varphi + \imath \sin \varphi] \neq 0$, then using the addition formulas for the sine and cosine functions, one has

$$zw = (r\rho)[\cos(\theta + \varphi) + \imath \sin(\theta + \varphi)] .$$

---

[3]The number π will be defined rigorously in Definition 3.29. Trigonometric functions will be introduced in the next chapter. Hence, for the moment, polar coordinates should not be used in proofs.

This polar form of the multiplication formula shows that complex multiplication involves real multiplication of the moduli and addition of the arguments, giving a geometric interpretation of how the operation of multiplication acts on vectors given in polar coordinates.

In particular, it follows that if $n \in \mathbb{Z}$ and $z = r\,(\cos\theta + \imath\,\sin\theta)$ is a nonzero complex number, then

$$z^n = r^n[\cos n\theta + \imath\,\sin n\theta] \;.$$

Therefore, for n in $\mathbb{Z}_{>0}$, each nonzero complex number z has (precisely) n n-th roots given by

$$r^{\frac{1}{n}}\left[\cos\left(\frac{\theta + 2\pi k}{n}\right) + \imath\,\sin\left(\frac{\theta + 2\pi k}{n}\right)\right]\;,$$

with $k = 0, 1, \ldots, n-1$.

The formula $d(z, w) = |z - w|$, for z and $w \in \mathbb{C}$, defines a **metric for** $\mathbb{C}$ that agrees with the Euclidean metric on $\mathbb{R}^2$ (under the linear representation of the complex plane).

DEFINITION 2.1. We say that a sequence (indexed by $n \in \mathbb{Z}_{>0}$) $\{z_n\}$ of complex numbers *converges* to $\alpha \in \mathbb{C}$ if given $\epsilon > 0$, there exists an $N \in \mathbb{Z}_{>0}$ such that $|z_n - \alpha| < \epsilon$ for all $n > N$; in this case we write

$$\lim_{n\to\infty} z_n = \alpha \;.$$

A sequence $\{z_n\}$ of complex numbers is called *Cauchy* if given $\epsilon > 0$, there exists an $N \in \mathbb{Z}_{>0}$ such that $|z_n - z_m| < \epsilon$ for all $n, m > N$.

THEOREM 2.2. *If $\{z_n\}$ and $\{w_n\}$ are Cauchy sequences of complex numbers, then*

(a) $\{z_n + \alpha\,w_n\}$ is Cauchy for all $\alpha \in \mathbb{C}$.
(b) $\{\bar{z}_n\}$ is Cauchy.
(c) $\{|z_n|\} \subset \mathbb{R}_{\geq 0}$ is Cauchy.

PROOF. (a) It suffices to assume that $\alpha \neq 0$. Given $\epsilon > 0$, choose N_1 such that $|z_n - z_m| < \frac{\epsilon}{2}$ for all $n, m > N_1$ and choose N_2 such that $|w_n - w_m| < \frac{\epsilon}{2|\alpha|}$ for all $n, m > N_2$. Choose $N = \max\{N_1, N_2\}$. Then for n and $m > N$, we have

$$|(z_n + \alpha\,w_n) - (z_m + \alpha\,w_m)| \leq |z_n - z_m| + |\alpha|\,|w_n - w_m| < \epsilon.$$

(b) $|\bar{z}_n - \bar{z}_m| = |z_n - z_m|$.

(c) Note that for all z and ζ in $\mathbb{C}$, we have

$$|z| = |z - \zeta + \zeta| \le |z - \zeta| + |\zeta|,$$

and hence, we conclude that

$$||z| - |\zeta|| \le |z - \zeta|.$$

Thus we have

$$||z_n| - |z_m|| \le |z_n - z_m|.$$

$\square$

REMARK 2.3. The above arguments mimic arguments in real analysis needed to establish the corresponding results for real sequences. We will, in the sequel, leave such routine arguments as exercises for the reader.

COROLLARY 2.4. $\{z_n\}$ *is a Cauchy sequence of complex numbers if and only if* $\{\Re z_n\}$ *and* $\{\Im z_n\}$ *are Cauchy sequences of real numbers.*

COROLLARY 2.5. $(\mathbb{C}, d)$ *is a complete metric space; that is, every Cauchy sequence of complex numbers converges to a complex number.*

PROOF. The metric on $\mathbb{C}$ restricts to the Euclidean metric on $\mathbb{R}$, which is complete. $\square$

DEFINITION 2.6. Let $A \subseteq \mathbb{C}$. Define

$$\|A\| = \{|z|\,;\ z \in A\} \subset \mathbb{R}_{\ge 0}.$$

We say that A is *bounded* if and only if $\|A\|$ is; that is, if there exists a positive real number M such that $|z| < M$ for all z in A.

DEFINITION 2.7. Let $\zeta \in \mathbb{C}$ and $\epsilon > 0$. The ϵ-*ball about* ζ is the set

$$U_\zeta(\epsilon) = U(\zeta, \epsilon) = \{z \in \mathbb{C}; |z - \zeta| < \epsilon\}.$$

PROPOSITION 2.8. *A subset A of $\mathbb{C}$ is bounded if and only if there exists a $\zeta \in \mathbb{C}$ and an $R > 0$ such that $A \subset U(\zeta, R)$.*

REMARK 2.9. A proof is omitted for one of three reasons (in addition to the reason described in Remark 2.3): Either it is trivial, it follows directly from results in real analysis, or it appears as an exercise at the end of the corresponding chapter.[4] The third possibility is always labeled; when standard results in real analysis are needed, there is some indication of what they are and where to find them. It should

---

[4]Exercises can be found at the end of each chapter and are numbered by chapter, so that Exercise 2.7 is to be found at the end of Chapter 2.

be clear from the context when the first possibility occurs. It is recommended that the reader check that he/she can supply an appropriate proof when none is given.

THEOREM 2.10 (**Bolzano–Weierstrass Theorem**). *Every infinite bounded set S in $\mathbb{C}$ has at least one limit point; that is, there exists at least one $a \in \mathbb{C}$ such that for all $\epsilon > 0$, $U(a, \epsilon)$ contains a point $z \neq a$, $z \in S$. Such an a is called a limit point of S.*

THEOREM 2.11. *A set $K \subset \mathbb{C}$ is compact if and only if it is closed and bounded.*

DEFINITION 2.12. Let f be a function defined on a set S in $\mathbb{C}$. We always (unless otherwise stated) assume that f is complex valued. Thus, f may be viewed as a map from S into $\mathbb{R}^2$ or $\mathbb{C}$ and as two real-valued functions defined on the set S. Let ζ be a limit point of S and $\alpha \in \mathbb{C}$. Then

$$\lim_{z \to \zeta} f(z) = \alpha$$

if and only if for all $\epsilon > 0$, there exists a $\delta > 0$ such that

$$|f(z) - \alpha| < \epsilon \text{ whenever } z \in S \text{ and } 0 < |z - \zeta| < \delta .$$

REMARK 2.13. In addition to the usual algebraic operations on pairs of functions $f : S \to \mathbb{C}$ and $g : S \to \mathbb{C}$ familiar from real analysis, such as $f + cg$ with $c \in \mathbb{C}$, fg, and $\frac{f}{g}$ (provided g does not vanish on S), we will consider other new functions constructed from a single function f. Among these

$$(\Re f)(z) = \Re f(z), \ (\Im f)(z) = \Im f(z), \ \overline{f}(z) = \overline{f(z)}, \ |f|\,(z) = |f(z)| .$$

THEOREM 2.14. *Let S be a set in $\mathbb{C}$, and let f and g be functions on S. Let ζ be a limit point of S. Then*

(a) $\lim_{z \to \zeta}(f + cg)(z) = \lim_{z \to \zeta} f(z) + c \lim_{z \to \zeta} g(z)$ *for all $c \in \mathbb{C}$,*
(b) $\lim_{z \to \zeta}(fg)(z) = \lim_{z \to \zeta} f(z) \lim_{z \to \zeta} g(z)$,
(c) $\lim_{z \to \zeta} |f|\,(z) = \left|\lim_{z \to \zeta} f(z)\right|$, *and*
(d) $\lim_{z \to \zeta} \overline{f}(z) = \overline{\lim_{z \to \zeta} f(z)}$.

REMARK 2.15. The usual interpretation of the above formulas is used here and in the rest of the book: The LHS[5] exists whenever the RHS exists, and then we have the stated equality.

[5]LHS (RHS) are standard abbreviations for left-(right)-hand side and will be used throughout this book.

COROLLARY 2.16. *Let* $u = \Re f$, $v = \Im f$ *(so that* $f(z) = u(z) + \imath v(z)$*) and* $\alpha \in \mathbb{C}$. *Then*

$$\lim_{z \to \zeta} f(z) = \alpha$$

if and only if

$$\lim_{z \to \zeta} u(z) = \Re\alpha \ and \ \lim_{z \to \zeta} v(z) = \Im\alpha \ .$$

DEFINITION 2.17. Let S be a subset of $\mathbb{C}$, $f : S \to \mathbb{C}$ and ζ a limit point of S, which is also in S. We say that f is *continuous at* ζ if $\lim_{z \to \zeta} f(z) = f(\zeta)$, that f is *continuous on* S if it is continuous at each ζ in S, and that f is *uniformly continuous on* S if and only if for all $\epsilon > 0$, there is a $\delta > 0$ such that

$$|f(z) - f(w)| < \epsilon \text{ for all } z \text{ and } w \text{ in } S \text{ with } |z - w| < \delta \ .$$

REMARK 2.18. Uniform continuity implies continuity.

THEOREM 2.19. *Let* f *and* g *be functions defined in appropriate sets; that is, sets where composition of these functions makes sense.*
(a) If f *is continuous at* ζ *and* $f(\zeta) \neq 0$, *then* $\frac{1}{f}$ *is defined in a neighborhood of* ζ *and is continuous at* ζ.
(b) If f *is continuous at* ζ *and* g *is continuous at* $f(\zeta)$, *then* $g \circ f$ *is continuous at* ζ.

THEOREM 2.20. *Let* $K \subset \mathbb{C}$ *be compact and* $f : K \to \mathbb{C}$ *be continuous. Then* f *is uniformly continuous on* K.

PROOF. A continuous mapping from a compact Hausdorff space to a metric space is uniformly continuous. ☐

DEFINITION 2.21. Given a sequence of functions $\{f_n\}$ all defined on the same set S in $\mathbb{C}$, we say that $\{f_n\}$ *converges uniformly* to a function f on S if for all $\epsilon > 0$, there exists an $N \in \mathbb{Z}_{>0}$ such that

$$|f(z) - f_n(z)| < \epsilon \text{ for all } z \in S \text{ and all } n > N \ .$$

REMARK 2.22. $\{f_n\}$ converges uniformly on S if and only if for all $\epsilon > 0$, there exists an $N \in \mathbb{Z}_{>0}$ such that

$$|f_n(z) - f_m(z)| < \epsilon \text{ for all } z \in S \text{ and all } n \text{ and } m > N \ .$$

THEOREM 2.23. *Let* $\{f_n\}$ *be a sequence of functions on* S. *If*
(1) $\{f_n\}$ *converges uniformly on* S, *and*
(2) *each* f_n *is continuous on* S,
  then the function f *defined by*

$$f(z) = \lim_{n \to \infty} f_n(z), \ z \in S \ ,$$

is continuous on S and $\{f_n\}$ converges uniformly to f on S.

PROOF. We start with two points z and ζ in S. Then for each n,

$$|f(z) - f(\zeta)| \leq |f(z) - f_n(z)| + |f_n(z) - f_n(\zeta)| + |f_n(\zeta) - f(\zeta)| .$$

Fix z and $\epsilon > 0$. By (1), the first and third term on the right-hand side are less than $\frac{\epsilon}{3}$ for n large. Fix n. By (2), the second term is less than $\frac{\epsilon}{3}$ as soon as ζ is close enough to z.   □

DEFINITION 2.24. A *domain* or *region* in $\mathbb{C}$ is a subset of $\mathbb{C}$ that is open and connected.

REMARK 2.25. Note that a domain in $\mathbb{C}$ could also be defined as an open arcwise-connected subset of $\mathbb{C}$.

2.2. Differentiability and holomorphic mappings

The definition of the derivative of a complex-valued function of a complex variable mimics that for the derivative of a real-valued function of a real variable. We shall see that the properties of the two classes of functions are quite different.

DEFINITION 2.26. Let f be a function defined in some ball about $\zeta \in \mathbb{C}$. Assume $h \in \mathbb{C}$. We say that f is *(complex) differentiable at ζ* if and only if

$$\lim_{h \to 0} \frac{f(\zeta + h) - f(\zeta)}{h}$$

exists. In this case the limit is denoted by

$$f'(\zeta), \quad \frac{df}{dz}(\zeta), \quad \frac{df}{dz}\bigg|_{z=\zeta}, \quad (Df)(\zeta) ,$$

and is called the *derivative of f at ζ*.

REMARK 2.27. (1) It is important that h is an arbitrary *complex* number (of small nonzero modulus) in the above definition.
(2) If f is differentiable at ζ, then f is continuous at ζ.

NOTATION 2.28. If the function f is differentiable on a domain D (that is, at each point of D), then it defines a function $f' : D \to \mathbb{C}$.
Thus for every $n \in \mathbb{Z}_{\geq 0}$, we can define inductively $f^{(n)}$, the n-th derivative of f, as follows:
$f^{(0)} = f$, and if $f^{(n)}$ is defined for $n \geq 0$, then we set $f^{(n+1)} = \left(f^{(n)}\right)'$ whenever the appropriate limits exist.
It is customary to abbreviate $f^{(2)}$ and $f^{(3)}$ by f'' and f''', respectively.

DEFINITION 2.29. Let f be a function defined in a neighborhood of $\zeta \in \mathbb{C}$. Then f is *holomorphic* or *analytic at* ζ if it is differentiable in a neighborhood (perhaps smaller) of ζ. A function defined on an (open) set U is *holomorphic* or *analytic on* U if it is holomorphic at each point of U.

A function f is *anti-holomorphic* if $\overline{f}$ is holomorphic.

The usual rules of differentiation hold. Let f and g be functions defined in a neighborhood of $\zeta \in \mathbb{C}$, let k be a function defined in a neighborhood of $f(\zeta)$, and let $c \in \mathbb{C}$. Then (recall Remark 2.15)

(a) $(f + cg)'(\zeta) = f'(\zeta) + cg'(\zeta)$,

(b) $(fg)'(\zeta) = f(\zeta)g'(\zeta) + f'(\zeta)g(\zeta)$,

(c) $(k \circ f)'(\zeta) = k'(f(\zeta))f'(\zeta)$,

(d) $\left(\dfrac{1}{f}\right)'(\zeta) = -\dfrac{f'(\zeta)}{f(\zeta)^2}$ provided $f(\zeta) \neq 0$, and

(e) for $f(z) = z^n$ $(n \in \mathbb{Z})$, $f'(z) = n\, z^{n-1}$ (for $n < 0$, $z \neq 0$).

DEFINITION 2.30. A function is called *entire* if it is holomorphic on $\mathbb{C}$.

EXAMPLE 2.31. (1) Every polynomial (in one complex variable) is entire.

(2) A rational function $R = \frac{P}{Q}$, where P and Q are polynomials with Q not the zero polynomial, is holomorphic on $\mathbb{C} - \{\text{zeros of } Q\}$. The polynomial Q has only finitely many zeros (the number of zeros, properly counted, equals the *degree* of Q; see Exercise 3.17).

(3) A special case of Example 2.31.2 is $R(z) = \frac{az+b}{cz+d}$ with a, b, c, and d fixed complex numbers satisfying $ad - bc = 1$. These rational functions are called *fractional linear transformations* or *Möbius transformations*, and they will be studied in detail in Section 8.1.

Convention. Whenever we write $z = x + \imath y$ for variables and $f = u + \imath v$ for functions, then we automatically mean that $x = \Re z$, $y = \Im z$, $u = \Re f$, and $v = \Im f$. We write $u = u(x, y)$ and $v = v(x, y)$ as well as $u = u(z)$ and $v = v(z)$.

THEOREM 2.32. *If $f = u + \imath v$ is differentiable at $c = a + \imath b$, then u and v have partial derivatives with respect to x and y at c, and these satisfy the Cauchy–Riemann equations (to be abbreviated CR):*

$$u_x(a, b) = v_y(a, b), \quad u_y(a, b) = -v_x(a, b). \tag{CR}$$

PROOF. First take $h = \alpha$ and then $h = \imath\beta$ (with α and β real) in the definition of differentiability (2.26) and compute

$$f'(c) = u_x(a,b) + \imath\, v_x(a,b) = -\imath\, u_y(a,b) + v_y(a,b) \ .$$

$\square$

We will also use the obvious notation $f_x = u_x + \imath v_x$ and $f_y = u_y + \imath v_y$.

REMARK 2.33. The CR equations are not sufficient for differentiability. To see this, define

$$f(z) = \begin{cases} z^5 \, |z|^{-4} & \text{for } z \neq 0, \\ 0 & \text{for } z = 0. \end{cases}$$

The function f is continuous on $\mathbb{C}$ and its real and imaginary parts satisfy the Cauchy–Riemann equations at $z = 0$, but it is not differentiable at $z = 0$. For α real and nonzero we have $\frac{f(\alpha)}{\alpha} = 1$, and for β real and nonzero we have $\frac{f(\imath\beta)}{\imath\beta} = 1$. Hence the CR equations are satisfied. Thus if the CR equations implied differentiability, we would conclude that $f'(0) = 1$. Now take $h = (1 + \imath)\gamma$ with γ real and nonzero and observe that $\frac{f(h)}{h} = -1$ so that $f'(0)$ would be -1.

In Exercise 2.8, we introduce the complex partial derivatives f_z and $f_{\bar{z}}$ of C^1-complex-valued functions[6] f defined on a region in the complex plane. These partials not only simplify the notation: For example, the two equations given in (CR) are written as the single equation

$$f_{\bar{z}} = 0 \,, \qquad\qquad \text{(CR complex)}$$

but they allow us to produce more concise arguments (and as we shall see later prettier formulas), as illustrated in the proof of the lemma below. We also use the notation $\frac{\partial f}{\partial \bar{z}}$ interchangeably with $f_{\bar{z}}$.

LEMMA 2.34. If f is a C^1-complex-valued function defined in a neighborhood of $c \in \mathbb{C}$, then for $z \in \mathbb{C}$ with $|z - c|$ small,

$$f(z) - f(c) = (z - c)f_z(c) + \overline{(z - c)}f_{\bar{z}}(c) + |z - c|\, \varepsilon(z, c), \qquad (2.1)$$

where $\varepsilon(z, c)$ is a complex-valued function of z and c with

$$\lim_{z \to c} \varepsilon(z, c) = 0 \ .$$

---

[6]C^1-complex-valued functions may be defined as functions whose real and imaginary parts have continuous first partial derivatives.

PROOF. As usual we write $z = x + \imath y$, $c = a + \imath b$, and $f = u + \imath v$ and abbreviate $\triangle u = u(z) - u(c)$, $\triangle x = x - a$, $\triangle y = y - b$, and $\triangle z = z - c = \triangle x + \imath \triangle y$.

By hypothesis, the real-valued function u has continuous first partial derivatives defined in a neighborhood of c, and we can define ε_1 by

$$\varepsilon_1(z, c) = \frac{\triangle u - u_x(c)\triangle x - u_y(c)\triangle y}{|\triangle z|}.$$

We show that

$$\lim_{z \to c} \varepsilon_1(z, c) = 0 . \qquad (2.2)$$

If we rewrite $\triangle u$ as

$$\triangle u = [u(x, y) - u(x, b)] + [u(x, b) - u(a, b)] ,$$

it follows from the (real) mean value theorem that

$$\text{RHS} = u_y(x, y_0)\triangle y + u_x(x_0, b)\triangle x ,$$

where y_0 is between y and b and x_0 is between x and a. Thus

$$\varepsilon_1(z, c) = \frac{[u_y(x, y_0) - u_y(a, b)]\triangle y + [u_x(x_0, b) - u_x(a, b)]\triangle x}{|\triangle z|}.$$

Hence we see that

$$|\varepsilon_1(z, c)| \leq |u_y(x, y_0) - u_y(a, b)| + |u_x(x_0, b) - u_x(a, b)| .$$

Thus we have shown that

$$u(z) - u(c) = (x - a)u_x(a, b) + (y - b)u_y(a, b) + |z - c|\,\varepsilon_1(z, c) ,$$

with (2.2).

Similarly,

$$v(z) - v(c) = (x - a)v_x(a, b) + (y - b)v_y(a, b) + |z - c|\,\varepsilon_2(z, c) ,$$

with

$$\lim_{z \to c} \varepsilon_2(z, c) = 0. \qquad (2.3)$$

With obvious notational conventions,

$$\triangle f = \triangle u + \imath \triangle v =$$
$$(u_x(a, b) + \imath v_x(a, b))\triangle x + (u_y(a, b) + \imath v_y(a, b))\triangle y + |\triangle z|\,\varepsilon(z, c)$$
$$= \frac{\triangle z + \overline{\triangle z}}{2}f_x(c) + \imath\frac{\overline{\triangle z} - \triangle z}{2}f_y(c) + |\triangle z|\,\varepsilon(z, c)$$
$$= \triangle z f_z(c) + \overline{\triangle z}f_{\bar z}(c) + |\triangle z|\,\varepsilon(z, c).$$

Since $\varepsilon(z, c) = \varepsilon_1(z, c) + \imath\, \varepsilon_2(z, c)$, equalities (2.2) and (2.3) imply that

$$\lim_{z \to c} \varepsilon(z, c) = 0.$$

$\square$

THEOREM 2.35. *If the function f has continuous first partial derivatives in a neighborhood of c that satisfy the CR equations at c, then f is (complex) differentiable at c.*

PROOF. The theorem is an immediate consequence of (2.1) since in this case $f_{\bar{z}}(c) = 0$ and hence $f'(c) = f_z(c)$. $\square$

COROLLARY 2.36. *If the function f has continuous first partial derivatives in an open neighborhood U of $c \in \mathbb{C}$ and the CR equations hold at each point of U, then f is holomorphic at c (in fact on U).*

REMARK 2.37. The converse is also true. It will take us some time to prove it.

THEOREM 2.38. *If f is holomorphic and real valued on a domain D, then f is constant.*

PROOF. As usual we write $f = u + \imath v$; in this case $v = 0$. The CR equations say $u_x = v_y = 0$ and $u_y = -v_x = 0$. Thus u is constant. $\square$

THEOREM 2.39. *If f is holomorphic and $f' = 0$ on a domain D, then f is constant.*

PROOF. As above $f = u + \imath v$ and $f' = u_x + \imath v_x = 0$. The last equation together with the CR equations say $0 = u_x = v_y$ and $0 = v_x = -u_y$. Thus both u and v are constant. $\square$

Exercises

2.1. (a) Let $\{z_n\}$ be a sequence of complex numbers and assume

$$|z_n - z_m| < \frac{1}{1 + |n - m|}\,, \quad \text{for all } n \text{ and } m\;.$$

Show that the sequence converges.
Do you have enough information to evaluate $\lim_{n \to \infty} z_n$?
What more can you say about this sequence?

(b) Let $\{z_n\}$ be a sequence with $\lim_{n \to \infty} z_n = 0$, and let $\{w_n\}$ be a bounded sequence. Show that

$$\lim_{n \to \infty} w_n z_n = 0\;.$$

2.2. Verify the Cauchy–Riemann equations for the function $f(z) = z^3$ by splitting f into its real and imaginary parts.

2.3. Let $x = r\cos\theta$, $y = r\sin\theta$. Show that the Cauchy–Riemann equations in polar coordinates for $F = U + \imath V$, $(U = U(r,\theta), V = V(r,\theta))$ are

$$r\frac{\partial U}{\partial r} = \frac{\partial V}{\partial \theta} \text{ and } r\frac{\partial V}{\partial r} = -\frac{\partial U}{\partial \theta}.$$

Alternatively, one can write

$$rU_r = V_\theta, \text{ and } rV_r = -U_\theta.$$

2.4. Suppose $z = x + \imath y$. Define

$$f(z) = \frac{xy^2\,(x+\imath y)}{x^2 + y^4},$$

for $z \neq 0$, and $f(0) = 0$.
Show that

$$\lim \frac{f(z) - f(0)}{z} = 0$$

as $z \to 0$ along any straight line. Show that as $z \to 0$ along the curve $x = y^2$, the limit of the difference quotient is $\frac{1}{2}$, thus showing that $f'(0)$ does not exist.

2.5. Does there exist a holomorphic function f on $\mathbb{C}$ whose real part is

(a) $u\,(x,y) = e^x$?
(b) or $u\,(x,y) = e^x(x\cos y - y\sin y)$?

Justify your answer. That is, if yes, exhibit the holomorphic function(s); if not, prove it.

2.6. Prove the *Fundamental Theorem of Algebra*: If $a_0,\dots,a_{n-1}$ are complex numbers $(n \geq 1)$ and $p(z) = z^n + a_{n-1}z^{n-1} + \dots + a_0$, then there exists a number $z_0 \in \mathbb{C}$ such that $p(z_0) = 0$.
Hints:

(a) Show there is an $M > 0$ and an $R > 0$ so that $|p(z)| \geq M$ for $|z| \geq R$.
(b) Show next that there is a $z_0 \in \mathbb{C}$ such that

$$|p(z_0)| = \min\{|p(z)|\,; z \in \mathbb{C}\}.$$

(c) By the change of variable $p(z + z_0) = g(z)$, it suffices to show that $g(0) = 0$.

(d) Write $g(z) = \alpha + z^m(\beta + c_1 z + \ldots + c_{n-m} z^{n-m})$ with $\beta \neq 0$. Choose γ such that

$$\gamma^m = -\frac{\alpha}{\beta}.$$

If $\alpha \neq 0$, obtain the contradiction $|g(\gamma z)| < |\alpha|$ for some z.

Note. We will later have a simpler proof of this theorem using results from complex analysis. See Theorem 5.15. See also the April 2006 issue of *The American Mathematical Monthly* for still other proofs of this fundamental result.

2.7. Using the Fundamental Theorem of Algebra stated in Exercise 2.6, prove *Frobenius Theorem*: If F is a field containing the reals and such that the dimension of F as a real vector space is finite, then either F is the reals or F is (isomorphic to) $\mathbb{C}$.
Hints:

(a) Assume $\dim_{\mathbb{R}} F = n > 1$. Show that for θ in $F - \mathbb{R}$ there exists a nonzero real polynomial p with leading coefficient 1 and such that $p(\theta) = 0$.
b) Show that there exist real numbers β and γ such that

$$\theta^2 - 2\beta\theta + \gamma = 0.$$

c) Show that there exists a positive real number δ such that $(\theta - \beta)^2 = -\delta^2$, and therefore,

$$\sigma = \frac{\theta - \beta}{\delta}$$

is an element of F satisfying $\sigma^2 = 1$.
d) The field

$$G = \mathbb{R}(\sigma) = \{x + y\sigma : x, y \in \mathbb{R}\} \subseteq F$$

is isomorphic to $\mathbb{C}$, so without loss of generality, assume $\sigma = \imath$ and $G = \mathbb{C}$. Conclude by showing that any element of F is the root of a complex polynomial with leading coefficient 1 and is therefore a complex number.

2.8. Let f be a complex-valued function defined on a region in the complex plane, and assume that both f_x and f_y exist in this region. Define:

$$f_z = \frac{1}{2}(f_x - \imath f_y)$$

and

$$f_{\bar z} = \frac{1}{2}(f_x + \imath f_y).$$

Show that for C^1-functions f,

$$f \text{ is holomorphic if and only if } f_{\bar{z}} = 0,$$

and that in this case $f_z = f'$.

2.9. Let R and Φ be two real-valued C^1-functions of a complex variable z. Show that $f = Re^{i\Phi}$ is holomorphic if and only if

$$\frac{R_{\bar{z}}}{R} = -i\Phi_{\bar{z}}.$$

2.10. Show that if f and g are C^1-functions, then the (complex) chain rule is expressed as follows (here $w = f(z)$ and g is viewed as a function of w):

$$(g \circ f)_z = g_w\, f_z + g_{\bar{w}}\, \overline{f}_z$$

and

$$(g \circ f)_{\bar{z}} = g_w\, f_{\bar{z}} + g_{\bar{w}}\, \overline{f}_{\bar{z}} .$$

2.11. Let p be a complex-valued polynomial of two real variables:

$$p(z) = \sum a_{ij} x^i y^j .$$

Write

$$p(z) = \sum_{j \geq 0} P_j(z)\bar{z}^j,$$

where each P_j is of the form $P_j(z) = \sum b_{ij} z^i$. Prove that p is an entire function if and only if

$$0 \equiv P_1 \equiv P_2 \equiv \dots.$$

2.12. (a) Given two points z_1, z_2 such that $|z_1| < 1$ and $|z_2| < 1$, show that for every point $z \neq 1$ in the closed triangle with vertices z_1, z_2 and 1,

$$\frac{|1 - z|}{1 - |z|} \leq K,$$

where K is a constant that depends only on z_1 and z_2.
(b) Determine the smallest value of K for $z_1 = \frac{1+i}{2}$ and $z_2 = \frac{1-i}{2}$.

2.13. Deduce the analogs of the CR equations for anti-holomorphic functions, in rectangular, polar, and complex coordinates.

2.14. Let D be an arbitrary (nonempty) open set in $\mathbb{C}$. Describe the class of complex-valued functions on D that are both holomorphic and anti-holomorphic.

2.15. (a) Every automorphism of the real field is the identity.
(b) Every continuous automorphism of the complex field is either the identity or the conjugation.

CHAPTER 3

Power Series

This chapter is devoted to an important method for constructing holomorphic functions. The tool is convergent power series. It is the basis for the introduction of new non-algebraic holomorphic functions, called elementary transcendental functions. It will turn out that all holomorphic functions are described (at least locally) by this tool. This will be proven in the next chapter.

The first section of this chapter is devoted to a discussion of elementary properties of the complex power series. Some material from real analysis, material not usually treated in books or courses on that subject, is studied. The concept of a convergent power series is extended from series with real coefficients to complex power series, and tests for convergence are established. In the second section, we show that convergent power series define holomorphic functions. The next section, Section 3.3, introduces important complex-valued functions of a complex variable, including the exponential function, the trigonometric functions, and the logarithm. This is followed by Sections 3.4 and 3.5, which describe an identity principle and introduce the new class of *meromorphic* functions: These functions are holomorphic on a domain except that they "assume the value ∞" (in a controlled way) at certain isolated points, known as the *poles* of the function. Meromorphic functions are defined locally as ratios of functions having power series expansions. It will hence follow subsequently that these are locally ratios of holomorphic functions. After some more work we will be able to replace "locally"by "globally." We develop the fundamental identity principle and its corollary, known as the *principle of analytic continuation*, for functions given by a power series, and we discuss the zeros and poles of a meromorphic function. The principle of analytic continuation is one of the most powerful results in complex function theory. Once we show that every holomorphic function is locally defined by a power series, we will see that the principle of analytic continuation says that a holomorphic function defined on an open connected set is remarkably rigid: Its behavior at a single point in the set determines its behavior at all other points of the set. Holomorphicity at a point is an extremely strong concept.

3.1. Complex power series

Let A be a subset of $\mathbb{C}$, and let $\{f_n\} = \{f_n\}_{n=0}^{\infty}$ be a sequence of functions defined on A (in the previous chapter, sequences were indexed by $\mathbb{Z}_{>0}$; for convenience, in this chapter, they will be indexed by $\mathbb{Z}_{\geq 0}$). We form the new sequence (known as a *series*) $\{S_N\}$, where

$$S_N(z) = \sum_{n=0}^{N} f_n(z), \text{ and the formal } \textit{infinite series} \sum_{n=0}^{\infty} f_n(z) .$$

The sequence of complex numbers $\{S_N(z)\}_{n=0}^{\infty}$ is also known as the *sequence of partial sums* associated with the infinite series $\sum_{n=0}^{\infty} f_n(z)$ at the point $z \in A$. When the indices of summation are clear from the context, we often omit them. For example, $\sum f_n(\zeta)$ usually means the infinite sum $\sum_{n=0}^{\infty} f_n(\zeta)$. Other similar abbreviations are used.

DEFINITION 3.1. i) We say that the infinite series $\sum_{n=0}^{\infty} f_n(z)$ *converges* at a point $\zeta \in A$ if $\{S_N(\zeta)\}$ converges. In this case, we write

$$\sum_{n=0}^{\infty} f_n(\zeta) = \lim_{n \to \infty} S_N(\zeta).$$

ii) We say that the infinite series $\sum_{n=0}^{\infty} f_n(z)$ *converges pointwise* in A if $\{S_N(\zeta)\}$ converges for every $\zeta \in A$.

iii) We say that the infinite series $\sum_{n=0}^{\infty} f_n(z)$ *converges absolutely* at a point $\zeta \in A$ if the infinite series $\sum_{n=0}^{\infty} |f_n(\zeta)|$ converges.

iv) We say that the infinite series $\sum_{n=0}^{\infty} f_n(z)$ *converges uniformly* in A if the sequence of partial sums $\{S_N(z)\}$ converges uniformly in A.

v) We say that the infinite series $\sum_{n=0}^{\infty} f_n(z)$ *converges normally* on a set $B \subset A$ if there exists a sequence of positive constants $\{M_n\}$ such that

(1) $|f_n(z)| \le M_n$ for all $z \in B$ and all n, and

(2) $\sum M_n < \infty$.

vi) We say that the infinite series $\sum_{n=0}^{\infty} f_n(z)$ *diverges* at a point of A if it does not converge at that point.

We speak of the *pointwise, uniform, absolute, or normal convergence* of a series as well as of the *divergence* of a series.

REMARK 3.2. The fact that the sequence $\{S_N(\zeta)\}$ converges if and only if $\{S_N(\zeta)\}$ is Cauchy allows us to rephrase conditions (i–iv) as if and only if statements when useful: (i) $\sum f_n(z)$ *converges* at a point $\zeta \in A$ if and only if $\{S_N(\zeta)\}$ is Cauchy; (ii) $\sum f_n(z)$ *converges pointwise* in A if and only if $\{S_N(\zeta)\}$ is Cauchy for every $\zeta \in A$; (iii) $\sum f_n(z)$ *converges absolutely* at a point $\zeta \in A$ if and only if the infinite series $\sum |f_n(\zeta)|$ converges; and (iv) $\sum f_n(z)$ *converges uniformly* in A if and only if the sequence of partial sums $\{S_N(z)\}$ converges uniformly in A.

REMARK 3.3. Many questions on convergence of complex sequences are reduced to the real case by the *trivial but important observation* that absolute convergence at a point implies convergence at that point.

REMARK 3.4. Two other such observations, both *trivial but important*, are that if an infinite series $\sum_{n=0}^{\infty} f_n(z)$ converges at a point ζ, then

$$\lim_{n \to \infty} f_n(\zeta) = 0 \text{ and } \lim_{n \to \infty} \sum_{k=n}^{\infty} f_k(\zeta) = 0.$$

Some relationships between some different types of convergence of a series are given in the following result.

Weierstrass M-test. Normal convergence implies uniform and absolute convergence.

PROOF. With notation as in the last definition, if $N_1 < N$ are positive integers, then

$$|S_N(z) - S_{N_1}(z)| \leq \sum_{n=N_1+1}^{N} M_n \text{ for all } z \in A$$

(needed for the uniform convergence argument), and

$$||S_N(z)| - |S_{N_1}(z)|| \leq \sum_{n=N_1+1}^{N} M_n \text{ for all } z \in A$$

(needed for the absolute convergence argument).

Given any $\epsilon > 0$, we can find a positive integer N_0 such that $N > N_1 > N_0$ implies $\sum_{n=N_1+1}^{N} M_n < \epsilon$. $\square$

We shall be mostly interested in series of the form $\sum_{n=0}^{\infty} a_n z^n$ with $a_n \in \mathbb{C}$ (these are known as *power* series) and the *associated* real-valued series $\sum_{n=0}^{\infty} |a_n| r^n$ where $r = |z|$. We define, for each $N \in \mathbb{Z}_{\geq 0}$,

$$S_N^*(r) = \sum_{n=0}^{N} |a_n| r^n \text{ for } r \in \mathbb{R}_{\geq 0},$$

and we observe that

$$S_{N+1}^*(r) \geq S_N^*(r) \text{ for all } N \in \mathbb{Z}_{\geq 0} \text{ and for all } r \in \mathbb{R}_{\geq 0}.$$

We refer to $S_N^*(r)$ as the *real partial sum at* r.

An *elementary but most important example* is provided by the *geometric series* with $r \in \mathbb{R}_{\geq 0}$:

$$1 + r + r^2 + \dots.$$

Note that $S_N(1) = N + 1$, and $S_N(r) = \dfrac{1 - r^{N+1}}{1 - r}$ for $0 \leq r < 1$, as well as for $r > 1$.

Thus $\sum_{n=0}^{\infty} r^n = \dfrac{1}{1 - r}$ if $0 \leq r < 1$ and $\sum r^n$ diverges if $r \geq 1$.

We now introduce two special cases of divergence of a sequence of real numbers. It will be useful to regard these sequences as convergent sequences with infinite limits.

DEFINITION 3.5. A sequence of real numbers $\{b_n\}$ *converges to* $+\infty$ if and only if for all $M > 0$ there exists an $N \in \mathbb{Z}_{>0}$ such that $b_n > M$ for all $n > N$. In this case we shall write $\lim_{n\to\infty} b_n = +\infty$. A similar definition applies to real sequences converging to $-\infty$.

With this notation either:

(a) $\lim_{N\to\infty} S_N^*(r)$ exists and is finite: That is, $\sum_{n=0}^{\infty} |a_n| r^n$ converges. In this case we write $\sum_{n=0}^{\infty} |a_n| r^n < +\infty$;

or

(b) $\lim_{N\to\infty} S_N^*(r) = +\infty$: That is, $\sum_{n=0}^{\infty} |a_n| r^n$ diverges (in the previous sense). In this case we write $\sum_{n=0}^{\infty} |a_n| r^n = +\infty$.

For real sequences, we have the **comparison test.** Let $0 \le a_n \le b_n$.

(a) If $\sum a_n = +\infty$, then $\sum b_n = +\infty$.
(b) If $\sum b_n < +\infty$, then $\sum a_n < +\infty$.

Abel's lemma. *Let* $0 < r < r_0$. *Assume there exists an* $M \in \mathbb{R}_{>0}$ *such that*

$$|a_n r_0^n| \le M \text{ for all } n \in \mathbb{Z}_{>0} .$$

Then the series $\sum a_n z^n$ *converges normally for all* z *with* $|z| \le r$. *In particular, it converges absolutely and uniformly for all* z *with* $|z| \le r$.

PROOF. For $|z| \le r$ we have

$$|a_n z^n| = |a_n| |z|^n \le |a_n| |r|^n = |a_n| \left(\frac{r}{r_0}\right)^n r_0^n \le M \left(\frac{r}{r_0}\right)^n .$$

Comparison with the geometric series [or an application of the Weierstrass M-test with $M_n = M(\frac{r}{r_0})^n$] shows the normal convergence. □

If S is any nonempty set of real numbers, then the least upper bound or *supremum* of S is denoted by $\sup S$ and the greatest lower bound or *infimum* is denoted by $\inf S$. The possibilities that $\sup S = +\infty$ or $\inf S = -\infty$ are allowed.

DEFINITION 3.6. The *radius of convergence* ρ of the power series $\sum a_n z^n$ is given by

$$\rho = \sup\{r \ge 0; \ \sum |a_n| r^n < +\infty\} .$$

Note that $0 \le \rho \le +\infty$. As a result of the next theorem it makes sense to define $\{z \in \mathbb{C}; |z| < \rho\}$ as the *disk of convergence* of the power series $\sum a_n z^n$.

THEOREM 3.7. *Let* $\sum a_n z^n$ *be a power series with radius of convergence* ρ.

    (a) *If* $0 < r < \rho$, *then* $\sum a_n z^n$ *converges normally, absolutely, and uniformly for* $|z| \leq r$.

    (b) *The series* $\sum a_n z^n$ *diverges for* $|z| > \rho$.

PROOF. (a) Choose r_0 with $r < r_0 < \rho$ such that $\sum |a_n| r_0^n < +\infty$. Thus there exists an $M > 0$ with $|a_n| r_0^n \leq M$ for all n in $\mathbb{Z}_{>0}$. Now apply Abel's lemma.

    (b) We claim that for $|z| > \rho$, the sequence $\{|a_n| |z|^n\}$ is not even bounded. Otherwise Abel's lemma (with $r_0 = |z|$) would guarantee the existence of an r with $\rho < r < |z|$ and $\sum |a_n| r^n < +\infty$. This contradicts the definition of ρ.   □

COROLLARY 3.8. *Let* $\sum a_n z^n$ *be a power series with radius of convergence* ρ. *Then the function defined by* $S(z) = \sum a_n z^n$ *is continuous for* $|z| < \rho$.

PROOF. It follows immediately from Theorems 2.23 and 3.7.   □

We now turn to the obvious and important question: *How do we compute* ρ?

To answer this question, we introduce the concepts of lim sup and lim inf.

DEFINITION 3.9. Let $\{u_n\}$ be a real sequence. We use $\equiv$ to indicate equivalent names [we also use this same notation with a different meaning in other places, such as $f \equiv 0$ or $f \equiv g$, to emphasize that these functions are (identically) equal] and define

$$\overline{\lim_n} \, u_n \equiv \limsup_n u_n \equiv \text{upper limit of } \{u_n\}$$

$$\equiv \text{ limit superior of } \{u_n\} = \lim_{p \to \infty} \sup_{n \geq p} \{u_n\} = \inf_p \sup_{n \geq p} \{u_n\},$$

and

$$\underline{\lim_n} \, u_n \equiv \liminf_n u_n \equiv \text{lower limit of } \{u_n\}$$

$$\equiv \text{ limit inferior of } \{u_n\} = \lim_{p \to \infty} \inf_{n \geq p} \{u_n\} = \sup_p \inf_{n \geq p} \{u_n\}.$$

Note that every real sequence has a limit superior as well as a limit inferior, which are either real numbers or $+\infty$ or $-\infty$.

Properties of limits superior and inferior. Let $\{u_n\}$ and $\{v_n\}$ be real sequences. Then:

(a) $\liminf_{n} u_n \leq \limsup_{n} u_n$.

(b) $\liminf_{n}(-u_n) = -\limsup_{n} u_n$.

(c) If $r > 0$, then

$$\liminf_{n}(ru_n) = r \liminf_{n} u_n, \text{ and } \limsup_{n}(ru_n) = r \limsup_{n} u_n.$$

(d) If $u_n \leq v_n$ for all n, then

$$\liminf_{n} u_n \leq \liminf_{n} v_n, \text{ and } \limsup_{n} u_n \leq \limsup_{n} v_n.$$

(e) $\lim_{n} u_n = L$ if and only if $\liminf_{n} u_n = L = \limsup_{n} u_n$
(L is allowed to be $\pm\infty$ in this context).

(f) $\liminf_{n}(u_n + v_n) \geq \liminf_{n} u_n + \liminf_{n} v_n$, and
$\limsup_{n}(u_n + v_n) \leq \limsup_{n} u_n + \limsup_{n} v_n$.

Remark and exercise. If f is a real-valued function of a complex variable, defined on a set S in $\mathbb{C}$ and if ζ is a limit point of S, then it is possible to define $\overline{\lim}_{z \to \zeta} f(z)$ and $\underline{\lim}_{z \to \zeta} f(z)$.

EXAMPLE 3.10. Let $u_n = \sin\left(\frac{n\pi}{2}\right), n = 0, 1, 2, \dots$. This is the sequence $\{0, 1, 0, -1, \dots\}$. Hence for all $p \in \mathbb{Z}_{\geq 0}$,

$$a_p = \sup_{n \geq p} u_n = 1 \text{ and thus } \limsup_{n} u_n = \lim_{p \to \infty} a_p = 1, \text{ and}$$

$$b_p = \inf_{n \geq p} u_n = -1 \text{ and thus } \liminf_{n} u_n = \lim_{p \to \infty} b_p = -1.$$

The **ratio test** will be well known to most readers (see Exercise 3.4). Suppose that $v_n > 0$ for all non-negative integers n.

(a) If $\lim_{n \to \infty} \frac{v_{n+1}}{v_n} = L < 1$, then $\sum v_n$ converges.

(b) If $\lim_{n \to \infty} \frac{v_{n+1}}{v_n} = L > 1$, then $\sum v_n$ diverges.

Perhaps less familiar is the **root test**. Suppose that $v_n \geq 0$ all n.

(a) If $\overline{\lim}_{n}(v_n)^{\frac{1}{n}} = L < 1$, then $\sum v_n$ converges.

(b) If $\overline{\lim}_{n}(v_n)^{\frac{1}{n}} = L > 1$, then $\sum v_n$ diverges.

PROOF. (a) Choose $\epsilon > 0$ such that $0 < L + \epsilon < 1$. There exists a $P \in \mathbb{Z}_{>0}$ such that

$$\sup_{n \geq p} \left\{ v_n^{\frac{1}{n}} \right\} < L + \epsilon \quad \text{for all } p \geq P \,.$$

Thus

$$v_n < (L + \epsilon)^n \quad \text{for all } n \geq P,$$

and comparison with the geometric series yields convergence.

(b) Suppose that $\sum v_n < +\infty$. Then $\lim_n v_n = 0$. Thus there exists a P in $\mathbb{Z}_{>0}$ such that $n \geq P$ implies that $v_n < 1$ for all $n \geq P$. Hence $(v_n)^{\frac{1}{n}} < 1$ for all $n \geq P$ and therefore $\overline{\lim_n}(v_n)^{\frac{1}{n}} \leq 1$. □

REMARK 3.11. The root test applies (with the same value of L) whenever the ratio test applies. However, the converse is not true. To see this take a sequence where the ratios are alternately $\frac{1}{2}$ and $\frac{1}{8}$. Then the root test will apply with $L = \frac{1}{4}$, but the sequence of ratios, obviously, will not converge.

EXAMPLE 3.12. Consider the sequence n^{-s} with s a positive real number. Both the ratio and the root test end up with $L = 1$. The series diverges (converges to $+\infty$) for $0 < s \leq 1$ and converges for $s > 1$.

We now return to the study of the complex power series $\sum_{n=0}^{\infty} a_n z^n$ and to the problem of computing the radius of convergence.

THEOREM 3.13. Let $\sum a_n z^n$ be a power series. Suppose that $a_n \neq 0$ for all n and that the limit $\lim_n \left| \dfrac{a_{n+1}}{a_n} \right| = L$ exists, with $0 \leq L \leq +\infty$.
Then the radius of convergence ρ of the power series $\sum_{n=0}^{\infty} a_n z^n$ is $\frac{1}{L}$; in other words,

$$\frac{1}{\rho} = \lim_{n \to \infty} \left| \frac{a_{n+1}}{a_n} \right| \,,$$

where $\dfrac{1}{+\infty}$ is to be understood as being $= 0$.

The hypotheses required for this result to hold are strong. As pointed out in Remark 3.11, the ratio test is stronger than the root test. The next result provides a way of computing the radius of convergence for *any* power series.

THEOREM 3.14 (**Hadamard**). *The radius of convergence ρ of the power series $\sum_{n=0}^{\infty} a_n z^n$ is given by*

$$\frac{1}{\rho} = \overline{\lim_n} |a_n|^{\frac{1}{n}} .$$

PROOF. Let $L = \overline{\lim_n} |a_n|^{\frac{1}{n}}$. Thus $\overline{\lim_n} |a_n r^n|^{\frac{1}{n}} = rL$ for all $r \geq 0$ and we conclude by the root test that the associated series $\sum |a_n| r^n$ converges for $0 \leq r < \frac{1}{L}$ and diverges for $r > \frac{1}{L}$. Thus $\rho = \frac{1}{L}$. □

LEMMA 3.15. *Let $\sum u_n$ and $\sum v_n$ be two absolutely convergent series. Define*

$$w_n = \sum_{p=0}^{n} u_p v_{n-p} .$$

Then $\sum w_n$ is absolutely convergent and $\sum w_n = (\sum u_n)(\sum v_n)$.

PROOF. Let $\alpha_p = \sum_{n \geq p} |u_n|$ and $\beta_p = \sum_{n \geq p} |v_n|$. Then

$$\lim_p \alpha_p = 0 = \lim_p \beta_p .$$

Also

$$\sum_{n=0}^{N} |w_n| \leq \sum_{n=0}^{\infty} |u_n| \sum_{n=0}^{\infty} |v_n| = \alpha_0 \beta_0 < +\infty,$$

and therefore, $\sum_{n=0}^{\infty} |w_n| < +\infty$. Thus we have proven the absolute convergence of the new series.

To show the required equality, choose m and n with $m \geq 2n$ and consider

$$\left| \sum_{k=0}^{m} w_k - \left(\sum_{k=0}^{n} u_k \right) \left(\sum_{k=0}^{n} v_k \right) \right| = \mathfrak{L}.$$

We have to show that $\mathfrak{L} \to 0$ as $n \to \infty$ (we already know each of the above series converges). We rewrite

$$\mathfrak{L} = \left| \sum_{k=0}^{m} \sum_{i=0}^{k} u_i v_{k-i} - \sum_{j=0}^{n} \sum_{i=0}^{n} u_i v_j \right| .$$

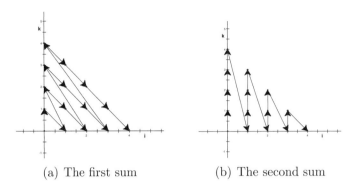

(a) The first sum   (b) The second sum

FIGURE 3.1. The (i, k) plane

By looking at the diagrams in the (i, k) plane shown in Figure 3.1, we see that

$$\sum_{k=0}^{m} \sum_{i=0}^{k} u_i v_{k-i} = \sum_{i=0}^{m} \sum_{k=i}^{m} u_i v_{k-i}$$

$$= \sum_{i=0}^{m} u_i \sum_{k=i}^{m} v_{k-i} = \sum_{i=0}^{m} u_i \sum_{j=0}^{m-i} v_j = \sum_{i=0}^{m} \sum_{j=0}^{m-i} u_i v_j \ .$$

Thus we can estimate

$$\mathfrak{L} = \left| \sum_{i=0}^{n} \left[\sum_{j=0}^{m-i} u_i v_j - \sum_{j=0}^{n} u_i v_j \right] + \sum_{i=n+1}^{m} \sum_{j=0}^{m-i} u_i v_j \right|$$

$$\leq \left| \sum_{i=0}^{n} \sum_{j=n+1}^{m-i} u_i v_j \right| + \left| \sum_{i=n+1}^{m} \sum_{j=0}^{m-i} u_i v_j \right|$$

$$\leq \sum_{i=0}^{\infty} \sum_{j=n+1}^{\infty} |u_i| \, |v_j| + \sum_{i=n+1}^{\infty} \sum_{j=0}^{\infty} |u_i| \, |v_j| = (\alpha_0 \beta_{n+1} + \beta_0 \alpha_{n+1}),$$

and the last expression approaches 0 as n goes to ∞. $\square$

3.2. More on power series

It is straightforward to prove the following (see Exercise 3.5).

THEOREM 3.16. *Suppose the power series $\sum a_n z^n$ and $\sum b_n z^n$ have radii of convergence $\geq \rho$. Then*

(a) *$\sum (a_n + b_n) z^n$ and $\sum c_n z^n$, where $c_n = \sum_{i=0}^{n} a_i b_{n-i}$, have radii of convergence $\geq \rho$, and*

(b) for $|z| < \rho$, we have

$$\sum (a_n + b_n) z^n = \sum a_n z^n + \sum b_n z^n$$

and

$$\sum c_n z^n = \left(\sum a_n z^n \right) \left(\sum b_n z^n \right).$$

THEOREM 3.17. *Suppose the power series $\sum_{n=0}^{\infty} a_n z^n$ has radius of convergence $\rho > 0$. For $|z| < \rho$, let*

$$S(z) = \sum_{n=0}^{\infty} a_n z^n$$

(note that this defines a continuous function by Corollary 3.8). Then the power series $\sum_{n=0}^{\infty} n a_n z^{n-1}$ has radius of convergence ρ and

$$S'(z) = \sum_{n=0}^{\infty} n a_n z^{n-1}, \quad \text{for all } |z| < \rho.$$

Thus a power series defines a function which is holomorphic (and C^{∞}) in its disk of convergence, and all its derivatives are holomorphic there.

PROOF. Let ρ' be the radius of convergence of $\sum_{n=0}^{\infty} n a_n z^{n-1}$. Then

$$\frac{1}{\rho'} = \overline{\lim_n}(n |a_n|)^{\frac{1}{n}} = \lim_n n^{\frac{1}{n}} \overline{\lim_n} |a_n|^{\frac{1}{n}} = \overline{\lim_n} |a_n|^{\frac{1}{n}} = \frac{1}{\rho},$$

where the first and last equality follow from Hadamard's theorem and the second one from Exercise 3.3.

Thus we can define a continuous function T on $|z| < \rho$ by

$$T(z) = \sum_{n=1}^{\infty} n a_n z^{n-1}.$$

Note that if A and $B \in \mathbb{C}$ and $n \in \mathbb{Z}_{>1}$, then

$$A^n - B^n = (A - B)(A^{n-1} + A^{n-2}B + \cdots + B^{n-1}).$$

Let us set, for $h \neq 0$ and $|h|$ sufficiently small,

$$f(h) = \left| \frac{S(z+h) - S(z)}{h} - T(z) \right| = \left| \sum_{n=2}^{\infty} a_n \left[\frac{(z+h)^n - z^n}{h} - n z^{n-1} \right] \right|$$

$$= \left| \sum_{n=2}^{\infty} a_n \left[(z+h)^{n-1} + z(z+h)^{n-2} + \ldots + z^{n-1} - n z^{n-1} \right] \right|.$$

For fixed z with $|z| < \rho$, we choose r so that $|z| < r < \rho$ and then we choose h with $|z + h| < r$. Under these restrictions

$$\left|(z + h)^{n-1} + z(z + h)^{n-2} + \ldots + z^{n-1} - nz^{n-1}\right|$$
$$\leq \left|(z + h)^{n-1} + z(z + h)^{n-2} + \ldots + z^{n-1}\right| + \left|nz^{n-1}\right| \leq 2nr^{n-1},$$

and thus

$$0 \leq f(h) \leq \sum_{n=2}^{\infty} 2\left|a_n\right| nr^{n-1} < +\infty.$$

Therefore given any $\epsilon > 0$, there exists a positive integer N with

$$\sum_{n=N}^{\infty} 2\left|a_n\right| nr^{n-1} < \frac{\epsilon}{2}.$$

Thus

$$0 \leq f(h) \leq \sum_{n=2}^{N-1} \left|a_n\right| \left|\frac{(z + h)^n - z^n}{h} - n\, z^{n-1}\right| + \sum_{n=N}^{\infty} 2\left|a_n\right| n\, r^{n-1}.$$

The first sum, being a finite sum, goes to 0 as $h \to 0$. Thus there exists a $\delta > 0$ such that $0 < |h| < \delta$ implies that the first term is at most $\frac{\epsilon}{2}$. Thus $0 \leq f(h) < \epsilon$ for $0 < |h| < \delta$. Hence $T(z) = S'(z)$. $\square$

REMARK 3.18. The theorem tells us that under certain circumstances we can interchange the order of computing limits: If the power series $\sum a_n z^n$ has a positive radius of convergence and if we let

$$\sum_{n=0}^{N} a_n z^n = S_N(z),$$

then the theorem states that

$$\lim_{h \to 0} \lim_{N \to \infty} \frac{S_N(z + h) - S_N(z)}{h} = \lim_{N \to \infty} \lim_{h \to 0} \frac{S_N(z + h) - S_N(z)}{h}.$$

COROLLARY 3.19. *If $S(z) = \sum_{n=0}^{\infty} a_n z^n$ for $|z| < \rho$, then for all $n \in \mathbb{Z}_{\geq 0}$ and all $|z| < \rho$,*

$$S^{(n)}(z) = \frac{d^n}{dz^n} S(z)$$

and

$$a_n = \frac{S^{(n)}(0)}{n!}.$$

PROOF. Induction on n shows that

$$S^{(n)}(z) = n! \, a_n + \frac{(n+1)!}{1!} \, a_{n+1} \, z + \dots \; .$$

□

The results obtained so far have provided information about the behavior of a power series inside its disk of convergence. Our next result deals with a point on the boundary of this disk.

THEOREM 3.20 (**Abel's limit theorem**). *Assume that the power series $\sum a_n z^n$ has finite radius of convergence $\rho > 0$. If $\sum a_n z_0^n$ converges for some z_0 with $|z_0| = \rho$, then $f(z) = \sum a_n z^n$ is defined for $\{|z| < \rho\} \cup \{z_0\}$ and we have*

$$\lim_{z \to z_0} f(z) = f(z_0),$$

as long as z approaches z_0 from inside the circle of convergence and

$$\frac{|z - z_0|}{\rho - |z|}$$

remains bounded.

PROOF. By the change of variable $w = \dfrac{z}{z_0}$ we may assume that $\rho = 1 = z_0$ (replace a_n by $a_n z_0^n$). Thus $\sum a_n$ converges to $f(1)$. By changing a_0 to $a_0 - f(1)$, we may assume that $f(1) = \sum a_n = 0$.

Therefore we are assuming that $|z| < 1$ (with $1 - |z|$ small) and that $\dfrac{|1 - z|}{1 - |z|} \leq M$ for some fixed $M > 0$. (Recall Exercise 2.12.) Let

$$s_n = a_0 + a_1 + \dots + a_n \, .$$

Then $\lim_n s_n = 0$ and

$$\begin{aligned}
S_n(z) &= a_0 + a_1 z + \dots + a_n z^n \\
&= s_0 + (s_1 - s_0)z + \dots + (s_n - s_{n-1})z^n \\
&= s_0(1 - z) + s_1(z - z^2) + \dots + s_{n-1}(z^{n-1} - z^n) + s_n z^n \\
&= (1 - z)(s_0 + s_1 z + \dots + s_{n-1} z^{n-1}) + s_n z^n.
\end{aligned}$$

Now

$$f(z) = \lim_{n \to \infty} S_n(z) = (1 - z) \sum_{n=0}^{\infty} s_n z^n \; .$$

Given $\epsilon > 0$, choose $N \in \mathbb{Z}_{>0}$ such that $|s_n| < \epsilon$ for $n > N$. Then

$$|f(z)| \leq |1 - z| \left(\left| \sum_{n=0}^{N} s_n z^n \right| + \sum_{n=N+1}^{\infty} |s_n| \, |z|^n \right)$$

$$\leq |1 - z| \left(\left| \sum_{n=0}^{N} s_n z^n \right| + \epsilon \frac{|z|^{N+1}}{1 - |z|} \right) \leq |1 - z| \left| \sum_{n=0}^{N} s_n z^n \right| + \epsilon M.$$

Thus we conclude that $\lim_{z \to 1} f(z) = 0$. $\qquad\qquad\qquad\qquad\square$

REMARK 3.21. Observe that we have not needed or used polar coordinates in our formal development thus far.

3.3. The exponential function, the logarithm function, and some complex trigonometric functions

In this section we use power series to develop several functions. We discover the exponential function by looking for functions that are solutions of the ordinary differential equation

$$f'(z) = f(z)$$

subject to the initial condition

$$f(0) = 1 \ .$$

3.3.1. The exponential function. We try to find such a solution defined by a power series

$$f(z) = a_0 + a_1 z + \cdots + a_n z^n + \ldots$$

that converges near $z = 0$. Then (by Theorem 3.17)

$$f'(z) = a_1 + 2a_2 z + \ldots + n a_n z^{n-1} + \ldots,$$

and thus we must have

$$a_0 = 1, \ a_1 = a_0 = 1, \ a_2 = \frac{1}{2} \cdot a_1 = \frac{1}{2!}, \ \ldots,$$

$$\ldots, \ a_{n+1} = \frac{1}{n+1} \cdot a_n = \frac{1}{(n+1)!} \quad \text{(by induction)}.$$

Hence

$$f(z) = 1 + z + \frac{z^2}{2!} + \ldots + \frac{z^n}{n!} + \ldots . \qquad\qquad (3.1)$$

Note that

$$\frac{1}{\rho} = \lim_n \left| \frac{a_{n+1}}{a_n} \right| = \lim_n \frac{n!}{(n+1)!} = 0.$$

Thus $\rho = +\infty$ and the power series (3.1) defines an entire transcen-

dental function. We write e^z and $\exp z$ for $f(z)$, and we call f the
exponential function.

A *transcendental* function is one that is not a rational function; that is, it is not the quotient of two polynomials. Exercise 3.9 shows that the exponential function is a transcendental function.

PROPOSITION 3.22. *Let $c \in \mathbb{C}$. Then the function $f(z) = c e^z$ is the unique power series and the unique entire function satisfying*

$$f'(z) = f(z) \quad and \quad f(0) = c. \tag{3.2}$$

PROOF. It is trivial that $z \mapsto c e^z$ satisfies (3.2) and is the unique power series to do so. We know that this is an entire function. We postpone the proof that this is the unique entire function that satisfies (3.2) until after we establish the next two propositions. □

PROPOSITION 3.23. $e^{z+\zeta} = e^z e^\zeta$ *for all z and ζ in $\mathbb{C}$.*

PROOF. Define $f(z) = e^{z+\zeta}$ with ζ fixed in $\mathbb{C}$. Then the function f has a power series expansion that converges for all $z \in \mathbb{C}$, $f'(z) = f(z)$ and $f(0) = e^\zeta$. Thus $f(z) = c e^z$ for some constant c, which must be $f(0) = e^\zeta$. □

PROPOSITION 3.24. *For all $z \in \mathbb{C}$,*

$$e^z e^{-z} = e^0 = 1.$$

Thus

$$e^z \neq 0 \text{ for all } z \in \mathbb{C}.$$

Conclusion of proof of Proposition 3.22. If $g(z)$ is any entire function satisfying (3.2), form the function $h(z) = \frac{g(z)}{e^z}$.

The rules for differentiation tell us that

$$h'(z) = \frac{g'(z)e^z - g(z)e^z}{e^{2z}} = 0 \text{ for all } z \in \mathbb{C}.$$

Thus, by Theorem 2.39 of the previous chapter, h is constant. □

PROPOSITION 3.25. $\overline{e^z} = e^{\bar{z}}$ *for all $z \in \mathbb{C}$.*

PROPOSITION 3.26. *Write $z = x + \imath y$. Then*

$$|e^{\imath y}|^2 = e^{\imath y} e^{-\imath y} = 1,$$

and thus

$$|e^z| = e^x.$$

The exponential function leads us immediately to the **complex trigonometric functions**.

3.3.2. The complex trigonometric functions. We define two entire functions by

$$\cos z = \frac{e^{iz} + e^{-iz}}{2} = 1 - \frac{z^2}{2!} + \frac{z^4}{4!} - \cdots \qquad (3.3)$$

and

$$\sin z = \frac{e^{iz} - e^{-iz}}{2i} = z - \frac{z^3}{3!} + \frac{z^5}{5!} - \cdots . \qquad (3.4)$$

It is then easy to verify the following familiar properties:

$$\cos z + i \sin z = e^{iz}, \ \cos^2 z + \sin^2 z = 1, \qquad (3.5)$$

$$\cos(-z) = \cos z, \ \ \sin(-z) = -\sin z,$$

$$\cos' z = -\sin z, \ \ \sin' z = \cos z. \qquad (3.6)$$

In the next subsection we will formally define π and then, after some calculations, we will be able to obtain from (3.5) the beautiful identity (connecting perhaps the four most interesting numbers in mathematics):

$$e^{\pi i} + 1 = 0.$$

REMARK 3.27. It should be observed that the functions sin and cos defined above agree for real values of the independent variable z with the familiar real-valued functions with the same names. The easiest way to conclude this is from the power series expansions of these functions at $z = 0$. Also note that $\sin z$ and $\cos z$ form a basis for the power series solutions to the ordinary differential equation

$$f''(z) + f(z) = 0.$$

Similarly, using either this last characterization of the sine and cosine functions or the additivity of the exponential function, one establishes that for all z and $\zeta \in \mathbb{C}$,

$$\cos(z + \zeta) = \cos z \cos \zeta - \sin z \sin \zeta$$

and

$$\sin(z + \zeta) = \sin z \cos \zeta + \cos z \sin \zeta .$$

3.3.3. The definition of π and the logarithm function. Our first task is to establish the **periodicity of e^z.** For $x \in \mathbb{R}$, $\sin x$ and $\cos x$ [as defined by (3.4) and (3.3) respectively] are real numbers. From $\sin^2 x + \cos^2 x = 1$, we conclude that

$$-1 \leq \cos x \leq 1 .$$

Integrating for $x \geq 0$ we have $\int_0^x \cos t\, dt \leq \int_0^x dt$, or $\sin x \leq x$. Also, for $x > 0$, we must have $\sin x < x$.[1] Equivalently, $-\sin x > -x$. Thus $\int_0^x (-\sin t)\, dt > \int_0^x (-t)\, dt$. We conclude that $\cos x - 1 > -\frac{x^2}{2}$ or $\cos x > 1 - \frac{x^2}{2}$. Repeating, we get two inequalities for $x \neq 0$:

$$\sin x > x - \frac{x^3}{6} \quad \text{and} \quad \cos x < 1 - \frac{x^2}{2} + \frac{x^4}{24} . \tag{3.7}$$

DEFINITION 3.28. Let f be a complex-valued function defined on $\mathbb{C}$, and let $c \in \mathbb{C}$, $c \neq 0$. We say that f has *period c* (and call f *periodic*) if and only if $f(z + c) = f(z)$ for all $z \in \mathbb{C}$.

The exponential function is periodic (that is, it has a period). Note that $e^{z+c} = e^z$ if and only if $e^c = 1$. Thus $c = \imath \omega$, with $\omega \in \mathbb{R}$. Traditionally ω (and not $\imath \omega$) is called a period of the exponential function. We want to determine the smallest such positive ω.

First note that $\cos 0 = 1$ (obvious). It follows from the inequalities (3.7) that

$$\cos \sqrt{3} < 1 - \frac{3}{2} + \frac{9}{24} = -\frac{1}{8} < 0,$$

and then continuity implies that there exists a $y_0 \in (0, \sqrt{3})$ such that $\cos y_0 = 0$.

But $\cos^2 y_0 + \sin^2 y_0 = 1$ implies that $\sin y_0 = \pm 1$, and thus $e^{\imath y_0} = \cos y_0 + \imath \sin y_0 = \pm \imath$ and $e^{4\imath y_0} = 1$. We conclude that $4y_0$ is a period of e^z. We claim that this is the smallest positive period and that any other period is an integral multiple of this one.

PROOF. If $0 < y < y_0 < \sqrt{3}$, then $y^2 < 3$ and $1 - \frac{y^2}{6} > \frac{1}{2}$; thus, $\sin y > y \left(1 - \frac{y^2}{6}\right) > \frac{y}{2} > 0$ and we conclude that $\cos y$ is strictly decreasing on $[0, y_0]$.

Since $\cos^2 x + \sin^2 x = 1$ and $\sin x > 0$ on $(0, y_0)$, we conclude that $\sin y$ is strictly increasing here. Thus, $\sin y < \sin y_0 = 1$.

---

[1] The function $x \mapsto x - \sin x$ is certainly nondecreasing on $[0, +\infty)$ since its derivative is the function $x \mapsto 1 - \cos x \geq 0$. The inequality $\sin x < x$ certainly holds for $x > 1$. If for some x_0 in $(0, 1]$ we would have $\sin x_0 = x_0$, then we would conclude from the Mean Value Theorem that for some $\tilde{x} \in (0, x_0)$, $\cos \tilde{x} = 1$, which leads to the contradiction $\sin \tilde{x} = 0$.

Now $0 < y < y_0$ implies that $0 < \sin y < 1$, which in turn implies that $e^{iy} \neq \pm 1, \pm i$ and therefore $e^{4iy} \neq 1$. Thus $\omega_0 = 4y_0$ is the smallest positive period.

If ω is an arbitrary period, then so is $|\omega|$ and there is an $n \in \mathbb{Z}_{>0}$ such that $n\omega_0 \leq |\omega| < (n+1)\omega_0$. If $n\omega_0 \neq |\omega|$, then $0 < (n+1)\omega_0 - |\omega|$ is a positive period less than ω_0. Since this is impossible by definition, $|\omega| = n\omega_0$. $\qquad\square$

DEFINITION 3.29. (**Definition of** π.) We define the real number π by $4y_0 = 2\pi$. Thus $e^z = 1$ if and only if $z = 2\pi i n$ with $n \in \mathbb{Z}$. Hence, according to tradition, 2π is the smallest positive period of the exponential function.

As in real analysis, the inverse to the exponential function should be a **logarithm**. We now turn to its definition. Since $e^x > 0$ for all $x \in \mathbb{R}$, e^x is strictly increasing on $\mathbb{R}$. Hence there exists an inverse function denoted by log (sometimes by ln):

$$e^x : \mathbb{R} \to (0, +\infty) , \qquad\qquad (3.8)$$
$$\log : (0, +\infty) \to \mathbb{R} ,$$

and we have the well-known properties

$$\log e^x = x \ \text{ for all } \ x \in \mathbb{R}, \ \text{ and } e^{\log x} = x \ \text{ for all } x \in \mathbb{R}_{>0} .$$

We know that $e^z \neq 0$ for all $z \in \mathbb{C}$; we thus can expect to define a complex logarithm. The problem is that the exponential function

$$e^z : \mathbb{C} \to \mathbb{C}_{\neq 0}$$

is not one-to-one. Let us write $z \neq 0$ in polar coordinates[2] $z = re^{i\theta}$ (this agrees with our previous way of writing polar coordinates). Here $r = |z|$ and $\theta = \arg z$. The argument of z is defined up to addition of $2\pi n$ with $n \in \mathbb{Z}$. We also define

$$\log z = \log |z| + i \arg z ;$$

it is a multivalued function[3] on $\mathbb{C}_{\neq 0}$.

The *principal branch* of $\arg z$, $\operatorname{Arg} z$, is restricted to lie in $(-\pi, \pi]$. It has a jump discontinuity on the negative real axis (it is not defined at 0). We define the *principal branch of the logarithm* by the formula

$$\operatorname{Log} z = \log |z| + i \operatorname{Arg} z .$$

---

[2]Having defined π, polar coordinates now rest on a solid foundation and can be used in proofs.

[3]Thus it is not a function.

It is a continuous function on $\mathbb{C} - (-\infty, 0]$; it is C^1 on this set.

Properties of the complex logarithm

1. $e^{\log z} = e^{\log|z| + i \arg z} = |z| \, e^{i \arg z} = z$ for all $z \in \mathbb{C}_{\neq 0}$.

2. Log z is holomorphic on $\mathbb{C} - (-\infty, 0]$, with $\dfrac{d}{dz} \text{Log } z = \dfrac{1}{z}$ there.

   PROOF. Write $z = r e^{i\theta}$ with $-\pi < \theta < \pi$. Thus Log $z = \log r + i\theta = u + iv$. Calculate $u_r = \frac{1}{r}$, $v_\theta = 1$, $u_\theta = 0$, and $v_r = 0$. Thus $r u_r = v_\theta$ and $r v_r = -u_\theta$. Hence Log z is C^1 and satisfies CR (see Exercise 2.3); thus it is holomorphic. We can now compute formally using the chain rule:
   $$e^{\text{Log } z} = z.$$

   Thus
   $$e^{\text{Log } z} \frac{d}{dz} \text{Log } z = 1,$$

   and we conclude that
   $$\frac{d}{dz} \text{Log } z = e^{-\text{Log } z} = \frac{1}{z}.$$
   $\square$

3. Log z_1 + Log z_2 = Log $z_1 z_2$ if $-\pi < \text{Arg } z_1 + \text{Arg } z_2 \le \pi$, and Log z_1 + Log $z_2 \neq$ Log $z_1 z_2$ otherwise.

DEFINITION 3.30. A continuous function f on a domain D not containing the origin is called a *branch of the logarithm* on D if for all $z \in D$, we have $e^{f(z)} = z$.

Later we will establish that under appropriate conditions on D a branch of the logarithm always exists.

THEOREM 3.31. *Let D be a domain in $\mathbb{C}$ with $0 \notin D$. Suppose f is a branch of the logarithm on D.*

Then g is a branch of the logarithm in D if and only if there is an $n \in \mathbb{Z}$ such that $g(z) = f(z) + 2\pi i n$ for all z in D.

PROOF. If $g = f + 2\pi i n$ with $n \in \mathbb{Z}$, then for all z in D, $e^{g(z)} = e^{f(z)} e^{2\pi i n} = z$.

For a proof of the converse, define
$$h(z) = \frac{f(z) - g(z)}{2\pi i}, \quad z \in D .$$

Then $e^{2\pi i h(z)} = e^{f(z)}e^{-g(z)} = z\dfrac{1}{z} = 1$. Thus for each $z \in D$, there is an $n \in \mathbb{Z}$ such that $h(z) = n$. Hence $h(D) \subset \mathbb{Z}$. Since D is connected, $h(D) = \{n\}$ for some fixed $n \in \mathbb{Z}$. $\qquad\qquad\qquad\qquad\square$

COROLLARY 3.32. *Every branch of the logarithm on a domain* D *is a holomorphic function on* D.

THEOREM 3.33. *For* $z \in \mathbb{C}$ *with* $|z| < 1$,

$$\mathrm{Log}(1+z) = \sum_{n=1}^{\infty}(-1)^{n-1}\frac{z^n}{n} = z - \frac{z^2}{2} + \frac{z^3}{3} + \dots \; .$$

PROOF. We first compute the radius of convergence of the given series using the ratio test: $\dfrac{1}{\rho} = \lim_{n}\left|\dfrac{n}{n+1}\right| = 1$. Thus

$$f(z) = \sum_{n=1}^{\infty}(-1)^{n-1}\frac{z^n}{n}$$

is holomorphic in $|z| < 1$. We calculate

$$f'(z) = 1 - z + z^2 + \dots = \frac{1}{1+z} \quad \text{for} \quad |z| < 1.$$

Let $g(z) = e^{f(z)}$; then $g'(z) = e^{f(z)}f'(z) = \dfrac{e^{f(z)}}{1+z}$ and

$$g''(z) = \frac{(1+z)e^{f(z)}f'(z) - e^{f(z)}}{(1+z)^2} = 0.$$

Thus $g'(z) = \alpha$, a constant, and

$$e^{f(z)} = \alpha(1+z).$$

Now $f(0) = 0$ tells us $\alpha = 1$. Thus $f(z)$ defines a branch of $\log(1+z)$. For $x \in (-1, 1)$, $f(x) \in \mathbb{R}$. Thus f is the principal branch of log; that is, $f(z) = \mathrm{Log}(1+z)$. $\qquad\qquad\qquad\qquad\square$

Complex exponentials are defined by $z^c = e^{c\log z}$ for $c \in \mathbb{C}$ and $z \in \mathbb{C}_{\neq 0}$ and the principal branch of z^c by $e^{c\,\mathrm{Log}\,z}$.

3.4. An identity principle

Holomorphic functions are remarkably rigid: If f is a holomorphic function that is defined on an open connected set D, then the knowledge of its behavior at a single point $a \in D$ is sufficient to describe precisely its properties [in particular, its value $f(b)$] at an arbitrary point $b \in D$. We now start on the exciting journey to establish this and other beautiful results.

DEFINITION 3.34. A function f defined in a neighborhood of $\zeta \in \mathbb{C}$ has *a power series expansion at* ζ if there exists an $r > 0$ such that

$$f(z) = \sum_{n=0}^{\infty} a_n (z - \zeta)^n \quad \text{for} \quad |z - \zeta| < r \le \rho,$$

where ρ is the radius of convergence of the power series (in the variable $w = z - \zeta$).

THEOREM 3.35. *Let f be a function defined in a neighborhood of $\zeta \in \mathbb{C}$ that has a power series expansion at ζ with radius of convergence ρ. Then*

(a) *f is holomorphic and C^∞ in a neighborhood of ζ.*

(b) *If g also has a power series expansion at ζ and if the product $f \cdot g$ is identically zero in a neighborhood of ζ, then either f or g is identically zero in some neighborhood of ζ.*

(c) *There exists a function h defined in a neighborhood of ζ that has a power series expansion at ζ, with the same radius of convergence ρ, such that $h' = f$. The function h is unique up to an additive constant.*

PROOF. Without loss of generality we assume $\zeta = 0$.

(a) Already verified, in Theorem 3.17.

(b) For some $r > 0$, we have

$$f(z) = \sum_{n=0}^{\infty} a_n z^n \text{ and } g(z) = \sum_{n=0}^{\infty} b_n z^n \text{ for } |z| < r.$$

Suppose that neither f nor g vanish identically in any neighborhood of $\zeta = 0$, and choose the smallest nonnegative integers N and M such that $a_N \ne 0$ and $b_M \ne 0$. We know that

$$(f \cdot g)(z) = \sum_{n=0}^{\infty} c_n z^n \text{ for } |z| < r ,$$

where

$$c_n = \sum_{p+q=n} a_p b_q.$$

Thus $c_{N+M} = a_N b_M \ne 0$ (note that $c_n = 0$ for $n = 0, 1, \ldots, N + M - 1$). But

$$c_{N+M} = \frac{1}{(N+M)!} \left(\frac{d^{N+M}(f \cdot g)}{dz^{N+M}} \right)(0) .$$

(c) Define $h(z) = \sum_{n=0}^{\infty} \frac{a_n}{n+1} z^{n+1}$. Then the radius of convergence ρ' of h satisfies

$$\frac{1}{\rho'} = \limsup_n \left| \frac{a_n}{n+1} \right|^{\frac{1}{n}} = \limsup_n |a_n|^{\frac{1}{n}} = \frac{1}{\rho} \ .$$

$\square$

The following lemma is a useful tool with significant applications beyond the immediate one.

LEMMA 3.36. *If $S(z) = \sum_{n=0}^{\infty} a_n z^n$ has radius of convergence $\rho > 0$, then for any $\zeta \in \mathbb{C}$ with $|\zeta| < \rho$, the power series*

$$\sum_{n=0}^{\infty} \frac{S^{(n)}(\zeta)}{n!} w^n$$

has radius of convergence $\geq \rho - |\zeta|$ and

$$S(z) = \sum_{n=0}^{\infty} \frac{S^{(n)}(\zeta)}{n!} (z - \zeta)^n \quad for \ |z - \zeta| < \rho - |\zeta| \ .$$

PROOF. Let us define $R = |\zeta| < \rho$ (see Figure 3.2). The argument consists of two steps:

(I) We show first that $\sum_{p=0}^{\infty} \frac{S^{(p)}(\zeta)}{p!} w^p$ is absolutely convergent for $|w| < \rho - R$.

We know that

$$S^{(p)}(\zeta) = \sum_{n=p}^{\infty} a_n n(n-1)(n-2)\ldots(n-p+1)\zeta^{n-p}$$

$$= \sum_{n=p}^{\infty} a_n \frac{n!}{(n-p)!}\zeta^{n-p} \ .$$

If $n - p = q$, we set $b_{p+q} = |a_n|$ and then

$$\left| S^{(p)}(\zeta) \right| \leq \sum_{q=0}^{\infty} b_{p+q} \frac{(p+q)!}{q!} R^q \ .$$

Take $r \in \mathbb{R}$ with $R < r < \rho$. Then

$$\sum_{p=0}^{\infty} \left| \frac{S^{(p)}(\zeta)}{p!} \right| (r - R)^p \leq \sum_{p,q} b_{p+q} \frac{(p+q)!}{p!\,q!} R^q (r - R)^p \qquad (3.9)$$

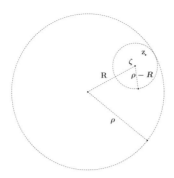

FIGURE 3.2. Radii of convergence

Making the change of variables $n = p + q$ in the RHS of (3.9) we continue to estimate the LHS.

$$\text{LHS} \le \sum_{n=0}^{\infty} b_n \sum_{p=0}^{n} \frac{n!}{p!(n-p)!} R^{n-p} (r-R)^p$$

$$= \sum_{n=0}^{\infty} b_n (r - R + R)^n < +\infty.$$

(II) We show next that

$$S(z) = \sum_{p=0}^{\infty} \frac{S^{(p)}(\zeta)}{p!} (z - \zeta)^p$$

$$= \sum_{p=0}^{\infty} \sum_{n=p}^{\infty} \frac{a_n n!}{p!(n-p)!} \zeta^{n-p} (z - \zeta)^p$$

for $|z - \zeta| < \rho - R$. We now know that this series converges absolutely. Hence we may rearrange the order of the terms and sums (see Exercise 3.13). The argument proceeds exactly as in part (I).

$\square$

EXAMPLE 3.37. $S(z) = \dfrac{1}{1-z}$ satisfies $S(z) = 1 + z + z^2 + \dots$ for $|z| < 1$. Here $\rho = 1$. The function S is defined on $\mathbb{C}_{\ne 1}$; however, the power series representation is valid only for $|z| < 1$.

Let us take $\zeta = -\frac{1}{2}$. Then $S^{(p)}(z) = p!\,(1-z)^{-1-p}$, and thus, $S^{(p)}\left(-\frac{1}{2}\right) = p!\left(\frac{2}{3}\right)^{1+p}$. Hence $\sum_{p=0}^{\infty}\left(\frac{2}{3}\right)^{1+p} w^p$ has radius of convergence $\rho' = \frac{3}{2}$. It follows that for all z satisfying $\left|z+\frac{1}{2}\right| < \frac{3}{2}$ we have

$$S(z) = \sum_{p=0}^{\infty}\left(\frac{2}{3}\right)^{1+p}\left(z+\frac{1}{2}\right)^{p}.$$

Note that $\rho - |\zeta| = \frac{1}{2} < \frac{3}{2}$. What we see in this example is not an accident as will soon become clear.

COROLLARY 3.38. *If $f(z) = \sum a_n z^n$ for $|z| < \rho$, then f has a power series expansion at each point ζ with $|\zeta| < \rho$.*

The next result is called an (perhaps, "the") *identity principle*; it provides necessary and sufficient conditions for a function that has a power series expansion at each point of a connected domain of definition to vanish identically. The principle is usually applied in the form given by Corollary 3.40.

THEOREM 3.39. *Let f be a function defined on a domain D. Assume that f has a power series expansion at each point of D, and let $\zeta \in D$.*

Then the following conditions are equivalent.

(a) $f^{(n)}(\zeta) = 0$ for $n = 0, 1, 2, \ldots$.

(b) $f \equiv 0$ in a neighborhood of ζ.

(c) There exists a sequence $\{z_n\}$ consisting of distinct points of D with $\lim_n z_n = \zeta$ and $f(z_n) = 0$ for each n.

(d) $f \equiv 0$ on D.

PROOF. It is obvious that $(d) \Rightarrow (b) \Rightarrow (c)$ and that $(a) \Leftrightarrow (b)$.

$(c) \Rightarrow (a)$: We know that $f(z) = \sum_{n=0}^{\infty} a_n(z - \zeta)^n$ for $|z - \zeta| < \rho$ for some $\rho > 0$. Furthermore, $a_0 = f(\zeta) = \lim_n f(z_n) = 0$.

Assume by induction that

$$0 = f(\zeta) = \ldots = f^{(n)}(\zeta)$$

for some integer $n \geq 0$. Then

$$f(z) = \sum_{p=n+1}^{\infty} a_p(z-\zeta)^p = (z-\zeta)^{n+1}\sum_{p=0}^{\infty} a_{n+1+p}(z-\zeta)^p = (z-\zeta)^{n+1}g(z).$$

Now, without loss of generality, we assume $z_n \neq \zeta$ for all n. The function g has a power series expansion at ζ. Obviously $g(z_n) = 0$ if and only if $f(z_n) = 0$. Thus $g(\zeta) = 0 = a_{n+1}$.

$(a) \Rightarrow (d)$: Let us define

$$D' = \{z \in D; f \equiv 0 \text{ in a neighborhood of } z\}.$$

The set D' is open in D because, as we already remarked, (a) is trivially equivalent to (b). Now

$$D' = \bigcap_{n=0}^{\infty} \{z \in D; f^{(n)}(z) = 0\}$$

is the intersection of a countable family of closed subsets of D and hence is closed in D. Since D' is not empty and D is connected, $D' = D$. $\square$

3.5. Zeros and poles

The most important and first practical consequence of the work of the last section is the next corollary. The results of Section 3.4 will also allow us to introduce an important class of functions; these are meromorphic functions taking values in the extended complex plane $\mathbb{C} \cup \{\infty\}$ rather than just $\mathbb{C}$.

COROLLARY 3.40 (**Principle of Analytic Continuation**). *Let D be a domain, and let f and g be functions on D having power series expansions at each point of D.*

If f and g agree on a sequence of distinct points in D with a limit point in D, or if they have identical power series expansions at a single point in D, then $f \equiv g$ (on D).

EXAMPLE 3.41. The exponential function e^z is the unique extension of e^x in the class of functions under study; that is, it is the unique function on $\mathbb{C}$ that has a power series expansions at each point and agrees with e^x at each point $x \in \mathbb{R}$.

COROLLARY 3.42. *If K is a compact subset of a domain D and f is a nonconstant function that has a power series expansion at each point of D, then f has finitely many zeros in K.*

DEFINITION 3.43. Let $\zeta \in \mathbb{C}$. Assume that

$$f(z) = \sum_{n=0}^{\infty} a_n(z - \zeta)^n \text{ for } |z - \zeta| < \rho \text{ for some } \rho > 0.$$

If f is not identically zero, then there exists an $N \in \mathbb{Z}_{\geq 0}$ such that

$$a_N \neq 0 \text{ and } a_n = 0 \text{ for all } n \text{ such that } 0 \leq n < N.$$

Thus

$$f(z) = (z - \zeta)^N \sum_{p=0}^{\infty} a_{N+p}(z - \zeta)^p = (z - \zeta)^N g(z),$$

with g having a power series expansion at ζ and $g(\zeta) \neq 0$. We define

$$N = \nu_\zeta(f) = \text{order (of the zero) of } f \text{ at } \zeta.$$

Note that $N \geq 0$, and $N = 0$ if and only if $f(\zeta) \neq 0$. If $N = 1$, then we say that f has a *simple* zero at ζ.

Notation. Let $\widehat{\mathbb{C}} = \mathbb{C} \cup \{\infty\}$ be the one point compactification of $\mathbb{C}$, usually called the *Riemann sphere*. It is diffeomorphic to the unit sphere in $\mathbb{R}^3$. See Exercise 3.18.

DEFINITION 3.44. (a) Let f be defined in a deleted neighborhood of $\zeta \in \mathbb{C}$ (see the Standard Notation summary). We say that

$$\lim_{z \to \zeta} f(z) = \infty$$

if for all $M > 0$, there exists a $\delta > 0$ such that

$$0 < |z - \zeta| < \delta \Rightarrow |f(z)| > M .$$

(b) Let $\alpha \in \widehat{\mathbb{C}}$, and let f be defined in $|z| > M$ for some $M > 0$ (we say that f is defined in a *deleted neighborhood* of ∞ in $\widehat{\mathbb{C}}$). We say

$$\lim_{z \to \infty} f(z) = \alpha$$

provided

$$\lim_{z \to 0} f\left(\frac{1}{z}\right) = \alpha.$$

(c) The above defines the concept of continuous maps between sets in $\widehat{\mathbb{C}}$.

(d) A function f is *holomorphic (has a power series expansion)* at ∞ if and only if $g(z) = f\left(\frac{1}{z}\right)$ is holomorphic (has a power series expansion) at $z = 0$.

DEFINITION 3.45. Let $U \subset \mathbb{C}$ be a neighborhood of a point ζ. A function f that is holomorphic in $U' = U - \{\zeta\}$, a deleted neighborhood of a point ζ, has a *removable singularity* at ζ if there is a holomorphic function in U that agrees with f on U'.

Let us consider two functions f and g having power series expansions at each point of a domain D in $\widehat{\mathbb{C}}$. Assume that neither function vanishes identically on D, and let $\zeta \in D \cap \mathbb{C}$. Let $F(z) = \dfrac{f(z)}{(z-\zeta)^{\nu_\zeta(f)}}$

and $G(z) = \frac{g(z)}{(z-\zeta)^{\nu_\zeta(g)}}$ for $z \in D$. Then the functions F and G have re-movable singularities at ζ, do not vanish there, and have power series expansions at each point of D. Furthermore

$$h(z) = \frac{f}{g}(z) = \frac{(z - \zeta)^{\nu_\zeta(f)} F(z)}{(z - \zeta)^{\nu_\zeta(g)} G(z)} \quad \text{for all } \zeta \in D.$$

Exactly three distinct possibilities exist for the behavior of $h(z)$ at ζ, which lead to the following definitions:

DEFINITION 3.46. (I) $\nu_\zeta(g) > \nu_\zeta(f)$. Then $h(\zeta) = \infty$ [this defines $h(\zeta)$, and the resulting function h is continuous at ζ]. In this case we say that h has a *pole of order* $\nu_\zeta(g) - \nu_\zeta(f)$ at ζ. If $\nu_\zeta(g) - \nu_\zeta(f) = 1$, we say that the pole is *simple*.

(II) $\nu_\zeta(g) = \nu_\zeta(f)$. The singularity of h at ζ is *removable*, and by definition, $h(\zeta) = \frac{F(\zeta)}{G(\zeta)}$.

(III) $\nu_\zeta(g) < \nu_\zeta(f)$. The singularity is again removable and $h(\zeta) = 0$.

In all cases we set $\nu_\zeta(h) = \nu_\zeta(f) - \nu_\zeta(g)$ and call it the *order* of h at ζ.

In cases (II) and (III) of the definition, h has a power series expansion at ζ as a consequence of the following

THEOREM 3.47. *If f has a power series expansion at ζ and $f(\zeta) \neq 0$, then $\frac{1}{f}$ also has a power series expansion at ζ.*

PROOF. Without loss of generality we assume $\zeta = 0$ and $f(0) = 1$. Thus

$$f(z) = \sum_{n=0}^{\infty} a_n z^n, \ a_0 = 1, \ \rho > 0 \ .$$

We want to find the reciprocal power series; that is, a power series $g(z) = \sum_{n=0}^{\infty} b_n z^n$ with a positive radius of convergence and such that

$$\left(\sum a_n z^n \right) \left(\sum b_n z^n \right) = 1 \ .$$

The LHS and the RHS are both power series, where the RHS is a power series expansion whose coefficients for $n > 1$ are all zero. Equating the first two coefficients, we obtain

$$a_0 b_0 = 1, \ b_0 = 1, \ a_1 b_0 + a_0 b_1 = 0,$$

and using the n-th coefficient of the power series when expanded for the LHS, for $n \geq 1$, we obtain

$$a_n b_0 + a_{n-1} b_1 + \ldots + a_0 b_n = 0 \ .$$

Thus by induction we define

$$b_n = -\sum_{j=0}^{n-1} b_j a_{n-j}.$$

Since $\rho > 0$, we have $\frac{1}{\rho} < +\infty$. Since $\limsup_n |a_n|^{\frac{1}{n}} = \frac{1}{\rho}$, there exists a $k > 0$ such that $|a_n| \le k^n$.

We will show by induction that $|b_n| \le 2^{n-1} k^n$ for all $n \ge 1$. For $n = 1$, we have $b_1 = -a_1$ and hence $|b_1| = |a_1| \le k$. Suppose the inequality holds for $j \le n$ for some $n \ge 1$. Then

$$|b_{n+1}| \le \sum_{j=0}^n |b_j| \, |a_{n+1-j}| = |a_{n+1}| + \sum_{j=1}^n |b_j| \, |a_{n+1-j}|$$

$$\le k^{n+1} + \sum_{j=1}^n 2^{j-1} k^j k^{n+1-j} = k^{n+1}(1 + 2^n - 1).$$

Thus there is a reciprocal series, with radius of convergence σ, satisfying

$$\frac{1}{\sigma} = \limsup_n |b_n|^{\frac{1}{n}} \le \lim_n (2^{1-\frac{1}{n}})k = 2\,k.$$

□

COROLLARY 3.48. *Let D be a domain in $\widehat{\mathbb{C}}$ and f a function on D. If f has a power series expansion at each point of D and $f(z) \ne 0$ for all $z \in D$, then $\frac{1}{f}$ has a power series expansion at each point of D.*

DEFINITION 3.49. For each domain $D \subseteq \mathbb{C} \cup \{\infty\} = \widehat{\mathbb{C}}$, we define $\mathbf{H}(D)$ as

$\{f : D \to \mathbb{C}; f$ has a power series expansion at each point of $D\}$.

The set $\mathbf{H}(D)$ is referred to as the set of holomorphic functions on D. We will see in Chapter 5 that this terminology is consistent with our earlier definition of a holomorphic function on D.

COROLLARY 3.50. *Assume that D is a domain in $\widehat{\mathbb{C}}$. The set $\mathbf{H}(D)$ is an integral domain and an algebra over $\mathbb{C}$. Its units are the functions that never vanish.*

DEFINITION 3.51. Let D be a domain in $\widehat{\mathbb{C}}$. A function $f : D \to \widehat{\mathbb{C}}$ is *meromorphic* if it is locally[4] the ratio of two functions having power series expansions (with the denominator not identically zero). The set of meromorphic functions on D is denoted by $\mathbf{M}(D)$.

---

[4]A property P is satisfied *locally* on an open set D if for each point $a \in D$, there exists a neighborhood $U \subset D$ of a such that P is satisfied in U.

COROLLARY 3.52. *Let D be a domain in $\widehat{\mathbb{C}}$, $f \in \mathbf{M}(D)$ and $\zeta \in D \cap \mathbb{C}$. Then there exist a connected neighborhood U of ζ in D, an integer n, and a unit $g \in \mathbf{H}(U)$ such that*

$$f(z) = (z - \zeta)^n\, g(z) \quad \text{for all } z \in U .$$

Note that $n = \nu_\zeta(f)$.

COROLLARY 3.53. *If D is a domain in $\widehat{\mathbb{C}}$, then the set $\mathbf{M}(D)$ is a field and an algebra over $\mathbb{C}$.*

We recall that, by our convention, $\mathbf{M}(D)_{\neq 0}$ is the set of meromorphic functions with the constant function 0 omitted, where $0(z) = 0$ for all z in D.

COROLLARY 3.54. *If D be a domain and $\zeta \in D$, then*

$$\nu_\zeta : \mathbf{M}(D)_{\neq 0} \to \mathbb{Z}$$

is a homomorphism.

Defining $\nu_\zeta(0) = +\infty$, we obtain

$$\nu_\zeta(f + g) \geq \min\{\nu_\zeta(f), \nu_\zeta(g)\} \quad \text{for all } f \text{ and } g \text{ in } \mathbf{M}(D);$$

that is, ν_ζ is a (discrete) valuation[5] (of rank one) on $\mathbf{M}(D)$.

REMARK 3.55. The converse statement also holds; it is non trivial and not established in this book.

The next corollary defines the term *Laurent series.*

COROLLARY 3.56. *If $f \in \mathbf{M}(D)_{\neq 0}$ and $\zeta \in D \cap \mathbb{C}$, then f has a Laurent series expansion at ζ; that is, there exist a $\mu \in \mathbb{Z}$ ($\mu = \nu_\zeta(f)$), a sequence of complex numbers $\{a_n\}_{n=\mu}^{\infty}$ with $a_\mu \neq 0$, and a deleted neighborhood U' of ζ such that*

$$f(z) = \sum_{n=\mu}^{\infty} a_n (z - \zeta)^n$$

for all $z \in U'$. The power series

$$\sum_{n=\max(0,\mu)}^{\infty} a_n (z - \zeta)^n$$

converges uniformly and absolutely on compact subsets of $U = U' \cup \{\zeta\}$.

---

[5]Standard, but not universal, terminology.

REMARK 3.57. If $\infty \in D$, then for all sufficiently large real numbers R the series representing f in $\{|z| > R\} \cup \{\infty\}$ has the form

$$f(z) = \sum_{n=\mu}^{\infty} a_n \left(\frac{1}{z}\right)^n .$$

COROLLARY 3.58. *If* $f \in \mathbf{M}(D)$, *then* $f' \in \mathbf{M}(D)$. *If in addition* $\nu_\varsigma(f) \neq 0$ *for* $\varsigma \in D$, *then*

$$\nu_\varsigma(f') = \nu_\varsigma(f) - 1 .$$

REMARK 3.59. Some care must be exercised in discussing singularities and singular values. ∞ is, of course, a singular value for holomorphic functions but not for meromorphic ones.

Exercises

3.1. Determine the radius of convergence of each of the following series:

$$\sum_{n=0}^{\infty} \frac{z^n}{n!}, \quad \sum_{n=1}^{\infty} \frac{z^n}{n}, \quad \sum_{n=0}^{\infty} n! z^n .$$

3.2. Prove that if $|a_n| \leq M$ for $n \geq 0$, then the power series $\sum_{n=0}^{\infty} a_n z^n$ has radius of convergence $\rho \geq 1$.

3.3. Under the hypothesis that $\{a_n\}$ and $\{b_n\}$ are positive sequences, prove that:

(a)
$$\overline{\lim_{n}} \, a_n b_n \leq \overline{\lim_{n}} \, a_n \, \overline{\lim_{n}} \, b_n,$$

provided the right side is not the indeterminate form $0 \times \infty$. Show by example that strict inequality may hold.

(b) If $\lim_n a_n$ exists, then the equality holds in (a) provided the right side is not indeterminate; that is, show that in this case,

$$\overline{\lim_{n}} \, a_n b_n = \lim_{n} a_n \, \overline{\lim_{n}} \, b_n.$$

3.4. Give a proof of the ratio test.

3.5. Give a proof of Theorem 3.16.

3.6. Find all the roots of $\cos z = 2$.

3.7. (a) Find $\Re(\sin z), \Im(\sin z), \Re(\cos z), \Im(\cos z)$.

(b) Write $z = x + \imath y$ and prove that

$$|\sin z|^2 = \sin^2 x + \sinh^2 y$$

and

$$|\cos z|^2 = \cos^2 x + \sinh^2 y,$$

where $\cosh z = \dfrac{e^z + e^{-z}}{2}$ and $\sinh z = \dfrac{e^z - e^{-z}}{2}$ are the *hyper-bolic trigonometric functions.*

(c) Derive the addition formulas for $\cosh(a + b)$ and $\sinh(a + b)$.

(d) Evaluate $D \sinh z$, $D \cosh z$, and $\cosh^2 z - \sinh^2 z$.

3.8. Prove, using power series, that $e^{-z} = \frac{1}{e^z}$.

3.9. Show that the exponential function is a transcendental function.

3.10. Is it always true that $\mathrm{Log}(e^z) = z$? Support your answer with either a proof or a counterexample.

3.11. (a) What are all the possible values of $\imath^{\imath}$?

(b) Let a and $b \in \mathbb{C}$ with $a \neq 0$. Find necessary and sufficient conditions for a^b to consist of infinitely many distinct values.

(c) Let n be a positive integer. Find necessary and sufficient conditions for a^b to consist of n distinct values.

3.12. (a) Show that both the sine and the cosine functions are periodic with period 2π.

(b) Show that $\sin z = 0$ if and only if $z = \pi n$ for some $n \in \mathbb{Z}$.

(c) Show that $\cos z = 0$ if and only if $z = \frac{\pi}{2}(2n + 1)$ for some $n \in \mathbb{Z}$.

3.13. Let $\{k_n\}$ be a sequence in which every positive integer appears once and only once.
Let $\sum a_n$ be a series. Putting $a'_n = a_{k_n}$, we say that $\sum a'_n$ is a *rearrangement* of $\sum a_n$.

(a) Let $\{a_n\}$ be a sequence of real numbers such that $\sum a_n$ converges but $\sum |a_n|$ does not. Let a be any real number. Show that there is a rearrangement $\sum a'_n$ of $\sum a_n$ such that $a = \sum a'_n$.

(b) Show that $\sum a_n$ converges absolutely if and only if every rearrangement converges to the same sum.

3.14. Let $\{a_n\}$ be a real sequence. Show that

$$\overline{\lim_n} \{a_n\} = \sup \left\{\alpha; \alpha = \lim_n b_n\right\},$$

with $\{b_n\}$ a convergent subsequence of $\{a_n\}$ and that

$$\varliminf_n a_n = \inf \left\{\alpha; \alpha = \lim_n b_n\right\}$$

with $\{b_n\}$ as above.

In this exercise a sequence $\{b_n\}$ with $\lim b_n = +\infty$ (similarly $-\infty$) is to be considered a convergent sequence.

3.15. (Only for those who know some algebra.) Show that $\mathbb{C}$ is the only nontrivial finite-dimensional commutative division algebra over $\mathbb{R}$.

3.16. Let $p(z) = a_n z^n + a_{n-1} z^{n-1} + \ldots + a_1 z + a_0$, $a_n \neq 0$, be a polynomial of degree $n \geq 1$. Consider p as a self-map of $\widehat{\mathbb{C}}$.

(a) Let $\alpha \in \widehat{\mathbb{C}}$. Show that there exists a $z \in \widehat{\mathbb{C}}$ such that $p(z) = \alpha$.[6]

(b) Let $z \in \widehat{\mathbb{C}}$ and $p(z) = \alpha \in \widehat{\mathbb{C}}$. Define appropriately $m_p(z)$, the *multiplicity* of α for p at z, so that you can prove: For all $\alpha \in \widehat{\mathbb{C}}$,

$$\sum_{z \in \widehat{\mathbb{C}};\, p(z)=\alpha} m_p(z) = n. \tag{3.10}$$

Note: The integer $\displaystyle\sum_{z \in \widehat{\mathbb{C}};\, p(z)=\alpha} m_p(z)$ is the *topological degree* of the map $p : \widehat{\mathbb{C}} \to \widehat{\mathbb{C}}$.

3.17. Let $p = \frac{P}{Q}$ be a nonconstant rational map. It involves no loss of generality to assume, as we do, that P and Q do not have any common zeros. View, as in the case of polynomials, p as a self-map of $\widehat{\mathbb{C}}$.

(a) Show that p is surjective.

(b) Define the concepts of multiplicity at a point and topological degree for the rational map p so that (3.10) holds.

3.18. The *unit sphere* (with center at 0) $S^2 \subseteq \mathbb{R}^3$ is defined by

$$S^2 = \{(\xi, \eta, \zeta) \in \mathbb{R}^3;\ \xi^2 + \eta^2 + \zeta^2 = 1\}.$$

---

[6]You may use, although other arguments are available, the Fundamental Theorem of Algebra which will be established in Chapter 5.

Show that *stereographic projection*

$$(\xi, \eta, \zeta) \mapsto \frac{\xi + i\eta}{1 - \zeta}$$

is a diffeomorphism from $S^2 - \{(0,0,1)\}$ onto $\mathbb{C}$ and that it extends to a diffeomorphism from S^2 onto $\widehat{\mathbb{C}}$ (that sends $(0,0,1)$ to ∞).

3.19. Justify the statement that stereographic projection takes circles to circles.

That is, a "circle" on S^2 is the intersection of a plane in $\mathbb{R}^3$ with S^2. Such a circle is a "maximal circle" if it is the intersection of S^2 with a plane through the point $(0,0,1)$ of $\mathbb{R}^3$.

Show that stereographic projection sets up a bijective correspondence between the set of "maximal circles" on S^2 and the set of "circles" through ∞ on $\widehat{\mathbb{C}}$, that is, straight lines in $\mathbb{C}$. Also show that stereographic projection sets up a bijective correspondence between the set of all circles on S^2 and what are called the *circles in* $\widehat{\mathbb{C}}$: the union of the set of all circles in $\mathbb{C}$ and the set of all straight lines in $\mathbb{C}$.

The circles in $\widehat{\mathbb{C}}$ will play an important role in Chapter 8.

3.20. Show that stereographic projection preserves angles.

3.21. Suppose the power series $\sum\limits_{-\infty}^{\infty} \alpha_j z^j$ and $\sum\limits_{-\infty}^{\infty} \beta_j z^j$ converge for $1 < |z| < 3$ and $2 < |z| < 4$, respectively, and that they have the same sum for $2 < |z| < 3$. Does this imply that $\alpha_j = \beta_j$ for all j?

3.22. Find all zeroes of $f(z) = 1 - \exp(\exp z)$.

3.23. Find the radius of convergence of the power series

$$\sum_{n=0}^{\infty} a_n z^n,$$

where $a_0 = 0$, $a_1 = 1$, and $a_n = a_{n-1} + a_{n-2}$ for all $n > 1$.

(Hint: Multiply the series by $z^2 + z - 1$.)

3.24. The formula

$$\tan z = \frac{\sin z}{\cos z}$$

defines a meromorphic function on $\mathbb{C}$. Show that it has simple poles at

$$z = (2k+1)\frac{\pi}{2} \text{ for every integer } k$$

and is holomorphic elsewhere. Show that tan maps $\mathbb{C}$ onto $\mathbb{C} \cup \{\infty\}$.

(a) Show that $\tan z = \tan \zeta$ if and only if there exists an integer k such that $\zeta - z = \pi k$.

(b) Show that $z \mapsto \tan z$ is a holomorphic one-to-one map of
$$\left\{ z \in \mathbb{C}; -\frac{\pi}{2} < \Re z < \frac{\pi}{2} \right\} \text{ onto } \mathbb{C} - \{(-\infty, -1) \cup (1, +\infty)\}$$
and of
$$\left\{ z \in \mathbb{C}; -\frac{\pi}{2} < \Re z \leq \frac{\pi}{2} \right\} \text{ onto } \mathbb{C} \cup \{\infty\}.^7$$

(c) Show that
$$\frac{d}{dz} \tan z = \frac{1}{\cos^2 z}.$$

3.25. One purpose of this exercise is to establish the beautiful formula (3.11).

Verify each of the following assertions and/or answer the questions:

(a) The series
$$\frac{1}{1 + z^2} = \sum_{k=0}^{\infty} (-1)^k z^{2k}$$
defines a holomorphic function on $|z| < 1$.

(b) Hence there is a holomorphic function
$$f(z) = \sum_{k=0}^{\infty} \frac{(-1)^k z^{2k+1}}{2k + 1}$$
on $|z| < 1$ such that
$$f(0) = 0 \text{ and } f'(z) = \frac{1}{1 + z^2}.$$

(c) Since tan is locally injective, there exists a multivalued inverse function arctan defined on $\mathbb{C} \cup \{\infty\}$ such that
$$\tan(\arctan z) = z \text{ for all } z \in \mathbb{C},$$
hence also $\tan(\arctan(z) + k\pi) = z$ for all $k \in \mathbb{Z}$ and all $z \in \mathbb{C}$. We can then define the *principal branch of* arctan, to be called Arctan, by requiring that
$$-\frac{\pi}{2} < \Re(\text{Arctan } z) \leq \frac{\pi}{2}.$$

(d) Show that
$$\arctan z = \frac{1}{2\imath} \log \frac{1 + \imath z}{1 - \imath z}.$$

---

[7] For this and the previous onto proof, you will need either some of the results of the next exercise or something like Rouché's theorem, which is proven in Chapter 6.

(e) Let $g(z) = f(\tan z)$. Show that $g'(z) = 1$ for all z in a domain D. Describe D.

(f) Conclude that $f(z) = \text{Arctan } z$ for $|z| < 1$.

(g) Why does the Taylor series for Arctan at the origin not converge in a bigger disk?

(h) Show that Arctan 1 is given by $\sum_{k=0}^{\infty} \frac{(-1)^k}{2k+1}$, thus justifying

$$\pi = 4 \sum_{k=0}^{\infty} \frac{(-1)^k}{2k+1}. \tag{3.11}$$

3.26. (l'Hopital's rule) Let f and g be two functions defined by convergent power series in a neighborhood of 0. Assume that $f(0) = 0 = g(0)$ and $g'(0) \neq 0$. Show that

$$\lim_{z \to 0} \frac{f(z)}{g(z)} = \frac{f'(0)}{g'(0)}.$$

CHAPTER 4

The Cauchy Theory–A Fundamental Theorem

As with the theory of differentiation for complex-valued functions of a complex variable, the integration theory of such functions begins by mimicking and extending results from the theory for real-valued functions of a real variable, but again the resulting theory is substantially different, more robust and elegant. Specifically, a curve or path γ in $\mathbb{C}$ is a continuous *function* from a closed interval in $\mathbb{R}$ to $\mathbb{C}$. Thus the restriction of a complex-valued function f on $\mathbb{C}$ to the range of a curve has real and imaginary parts that can be viewed as real-valued functions of a real variable and thus integrated on the interval.[1] Adding the integral of the real part to $\imath$ times the integral of the imaginary part defines a complex-valued integral of a complex function (that is, $\int f = \int \Re f + \imath \int \Im f$). In fact there are several useful ways to employ the ability to integrate a function of a real variable to define complex-valued integrals of a complex variable over certain paths. Among these integrals are those known as line integrals, complex line integrals, and integrals with respect to arc length. One can then use the integration theory of real variables to obtain an integration theory for complex functions along curves in $\mathbb{C}$. This extends to a more general theory, the Cauchy Theory, which constitutes a main portion of what we have called the Fundamental Theorem (Theorem 1.1). The integration theory depends not just on the integrated function being holomorphic but also on the topology of the curve over which the integration is being carried out and the topology of the domain in which the curve lies. In the simplest situation, Cauchy's theorem says that the integral of a holomorphic function over a simple closed curve lying in a simply connected domain is zero.[2]

In this chapter we lay the foundations for proving several of the equivalences in the Fundamental Theorem. Beginning in Section 4.1, we present a more or less self-contained treatment; our approach is

---

[1] When suitable conditions on $\Re f$ and $\Im f$ hold.

[2] A simple closed curve is one whose initial and end points coincide and has no other self-intersections. Being simply connected is a topological property of a domain (see Section 4.4).

through integration of closed forms over closed curves. Line integrals and differential forms are introduced in this section. In the next one we emphasize the difference between exact and closed (locally exact) one-forms. Integration of closed forms along continuous, not necessarily rectifiable, paths is discussed next. This is followed by a section on the winding number of a curve about a point. In the last section we treat Cauchy's theorem in its simplest format: that the integral of a holomorphic function over the boundary of a rectangle is zero. This is known as Cauchy's theorem for a rectangle.

Although the proofs of the main results become quite technical in places, the final results are simple to state. This simplicity gives them a certain elegance and compactness.

The two chapters that follow this one present the core of what we have termed the Fundamental Theorem, Theorem 1.1. In Chapter 5 we present key consequences of the initial Cauchy Theory (that is, Cauchy's theorem for a rectangle). This is followed by Chapter 6 where consequences related to holomorphic functions with isolated singularities are presented.

4.1. Line integrals and differential forms

We recall the definitions of the one-sided derivative for functions of a real variable.

DEFINITION 4.1. Let $[a, b]$ be a closed (finite) interval on $\mathbb{R}$ and let $g : [a, b] \to \mathbb{R}$ be a function. As in calculus, for $a \leq c < b$, we define

$$(D^+g)(c) = \lim_{\substack{h \to 0 \\ h > 0}} \frac{g(c + h) - g(c)}{h},$$

the *right-sided* derivative of g at c (whenever this limit exists).

Similarly, for $a < c \leq b$, we define

$$(D^-g)(c) = \lim_{\substack{h \to 0 \\ h < 0}} \frac{g(c + h) - g(c)}{h},$$

the *left-sided* derivative of g at c (whenever this limit exists), and for $a < c < b$, we define

$$g'(c) = (Dg)(c) = (D^+g)(c) = (D^-g)(c),$$

the *derivative* of g at c, whenever the last two limits exist and are equal.

We say that g is *differentiable* on $[a, b]$ if g' exists on (a, b) and $(D^+g)(a)$ and $(D^-g)(b)$ exist (these define $g'(a)$ and $g'(b)$, respectively); g is called *continuously differentiable* on $[a, b]$ if g' is continuous on

$[a, b]$; in which case, we will write $g \in \mathbf{C}^1_{\mathbb{R}}([a, b])$ or, equivalently, $g' \in \mathbf{C}^0_{\mathbb{R}}([a, b]) = \mathbf{C}_{\mathbb{R}}([a, b])$.

REMARK 4.2. The concepts we have been discussing are from real analysis; if $f : [a, b] \rightarrow \mathbb{C}$ is a complex-valued function, they apply to the two real-valued functions of a real variable given by $u = \Re f$ and $v = \Im f$, and we can set $f' = u' + \imath v'$. We abbreviate $\mathbf{C}^1_{\mathbb{C}}([a, b])$ and $\mathbf{C}^0_{\mathbb{C}}([a, b]))$ by $\mathbf{C}^1([a, b])$ and $\mathbf{C}^0([a, b]))$, respectively.

In what follows D is a domain in $\mathbb{C}$.

DEFINITION 4.3. A function $\gamma \in \mathbf{C}^1([a, b])$ with $\gamma([a, b]) \subset D \subseteq \mathbb{C} \cong \mathbb{R}^2$ will be called a *differentiable path* or *curve in D*, and we say that γ is *parameterized* by $[a, b]$.

We will write

$$\gamma(t) = (x(t), y(t)) = z(t) = x(t) + \imath y(t) \quad \text{for} \ a \le t \le b.$$

We denote the image or range of γ by range γ.

The curve is called *closed* if $\gamma(a) = \gamma(b)$. The closed curve γ is *simple* if γ is one-to-one except at the end points of the interval $[a, b]$; to be precise, if $\gamma(t_1) = \gamma(t_2)$ for $a \le t_1 < t_2 \le b$, then $a = t_1$ and $t_2 = b$.

DEFINITION 4.4. Let D be a domain in $\mathbb{C}$. A *differential (one-)form* ω *on D* is an expression

$$\omega = P \, dx + Q \, dy,$$

where $P = P(x, y)$ and $Q = Q(x, y)$ are continuous (complex-valued) functions on D, and dx and dy are symbols associated with the coordinate $z = x + \imath y$ and called the *differentials* of x and y, respectively.

If γ is a differentiable path in D and ω is a differential form on D, then we define *the line* or *path* or *contour integral of ω along γ*, by the formula

$$\int_\gamma \omega = \int_a^b [P(x(t), y(t)) \, x'(t) + Q(x(t), y(t)) \, y'(t)] dt$$

$$= \int_a^b (p_1(x(t), y(t)) \, x'(t) + q_1(x(t), y(t)) \, y'(t)) \, dt$$

$$+ \imath \int_a^b (p_2(x(t), y(t)) \, x'(t) + q_2(x(t), y(t)) \, y'(t)) \, dt,$$

where $P = p_1 + \imath p_2$ and $Q = q_1 + \imath q_2$.

REMARK 4.5. The above definition involves again only concepts from real analysis, even though the paths and functions involve the complex numbers.

Reparametrization: If $t : [\alpha, \beta] \to [a, b]$ is one-to-one, onto, and differentiable, and $\gamma : [a, b] \to \mathbb{C}$ is a differentiable path, then $\widetilde{\gamma} = \gamma \circ t$ is again a differentiable path, called a *reparametrization* of γ.

Since for any closed interval $[a, b]$ there exists a one-to-one, onto, and differentiable map $t : [0, 1] \to [a, b]$, we can always assume that a given path γ is parameterized by $[0, 1]$.

For all differential forms ω defined in a neighborhood of the range of the path γ and all reparametrizations $\widetilde{\gamma}$ of γ, the following equalities hold:

$$\int_{\widetilde{\gamma}} \omega = \int_{\gamma} \omega \; , \; \text{if} \; t'(u) \geq 0 \; \text{ for all } u \in [\alpha, \beta]$$

and

$$\int_{\widetilde{\gamma}} \omega = - \int_{\gamma} \omega \; , \; \text{if} \; t'(u) \leq 0 \; \text{ for all } u \in [\alpha, \beta] \, .$$

Note that there are no other possibilities for the sign of the derivative of t.

Subdivision of interval: Let $\gamma : [a, b] \to \mathbb{C}$ be a differentiable path and consider the *partition* of $[a, b]$ defined by

$$a = t_0 < t_1 < \ldots < t_{n+1} = b \, . \tag{4.1}$$

If

$$\gamma_j = \gamma|_{[t_j, t_{j+1}]} \; , \; \text{for} \; j = 0, \ldots, n, \tag{4.2}$$

then γ_j is a differentiable path and

$$\int_{\gamma} \omega = \sum_{j=0}^{n} \int_{\gamma_j} \omega. \tag{4.3}$$

For a set T contained in the domain of γ, $\gamma|_T$, of course, denotes the restriction of γ to T.

DEFINITION 4.6. Let $\gamma : [a, b] \to \mathbb{C}$ be a continuous path. We say that γ is a *piecewise differentiable path* (henceforth abbreviated *pdp*) if there exists a partition of $[a, b]$ of the form given in (4.1) such that each path defined by (4.2) is differentiable. Then we use (4.3) to define the path integral $\int_{\gamma} \omega$.

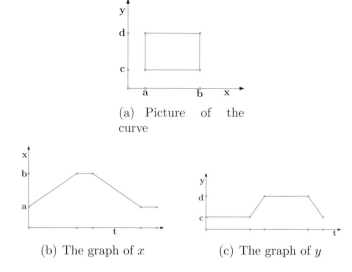

(a) Picture of the curve

(b) The graph of x

(c) The graph of y

FIGURE 4.1. Three figures

REMARK 4.7. The path integral is well defined (independent of the partition) and agrees with earlier definition for differentiable paths. The verification is left as an exercise.

REMARK 4.8. Three pictures in $\mathbb{R}^2$ are naturally associated with each path $\gamma = x + \imath y$: the picture of the curve and the graphs of the functions x and y. Figure 4.1 illustrates this with a curve whose image is the rectangle with vertices $(c, e), (d, e), (d, f)$, and (c, f).

LEMMA 4.9. *If D is a domain in $\mathbb{C}$, then any two points in D can be joined by a piecewise differentiable path in D.*

PROOF. Fix $\zeta \in D$, and let

$$E = \{z \in D; z \text{ can be joined to } \zeta \text{ by a pdp in } D\}.$$

Then E is open in D, $D - E$ is also open in D, and $\zeta \in E$. □

DEFINITION 4.10. Let D be a domain in $\mathbb{C}$.

(1) A function f on D is *of class C^p*, with $p \in \mathbb{Z}_{\geq 0}$, if f has partial derivatives (with respect to x and y) up to and including order p and these are continuous on D. It is of class C^∞ if it is of class C^p for all $p \in \mathbb{Z}_{\geq 0}$. The vector space of functions of class C^p on D is denoted by $\mathbf{C}^p(D)$.
(2) A differential form $\omega = P dx + Q dy$ is of class C^p if and only if P and Q are of that class.

(3) For a given function f, we have the (real) partial derivatives f_x and f_y as well as the formal (complex) partial derivatives f_z and $f_{\bar{z}}$ introduced in Exercise 2.8, where it was also shown that for a C^1-function $f = u + \imath v$, the Cauchy–Riemann equations hold for the pair u and v if and only if $f_{\bar{z}} = 0$.

  The four partial derivatives just described may be regarded as directional derivatives.

REMARK 4.11. At this point we recommend that all concepts and definitions that are formulated in terms of x and y be reformulated by the reader in terms of z and $\bar{z}$ (and vice versa).

(4) If f is a C^1-function on D, then we define df, the *total differential of* f, by either of the two equivalent formulas

$$df = f_x \, dx + f_y \, dy = f_z \, dz + f_{\bar{z}} \, d\bar{z},$$

where

$$dz = dx + \imath \, dy \quad \text{and} \quad d\bar{z} = dx - \imath \, dy.$$

Thus in addition to the differential operator d, we have two other important differential operators ∂ and $\bar{\partial}$ defined by

$$\partial f = f_z \, dz \quad \text{and} \quad \bar{\partial} f = f_{\bar{z}} \, d\bar{z}$$

as well as the formula

$$d = \partial + \bar{\partial}.$$

We have defined the three differential operators on spaces of C^1-functions. They can be also defined on spaces of C^1-differential forms, and it follows from these definitions that, for example, on C^2-functions the equality $d^2 = 0$ holds. We shall not need these extended definitions.

(5) A differential form w on an open set D is called *exact* if there exists a C^1-function F on D (called a *primitive for w*) such that

$$w = dF.$$

If D is connected, then a primitive (if it exists) is unique up to addition of a constant.

  By abuse of language we also say that a function F is a *primitive* for a function f if F is a primitive for the differential $w = f \, dz$.

(6) A differential form ω on D is called *closed* if it is locally exact; that is, if for each $a \in D$, there exists a neighborhood U of a such that $\omega_{|U}$ is exact.

4.2. The precise difference between closed and exact forms

Although the definitions of exact and closed are straightforward, as is the fact that every exact differential is closed, an intuitive sense of the difference between the two properties may not be immediately present itself. The reason is because these differences arise from the topology of the domain and the behavior of the differential form along certain paths. An example of a closed but not exact differential will be given in Example 4.24. We will see that on a disk the two properties are equivalent, but situations where they are not equivalent are especially significant.

To understand this difference, we study the pairing that associates the complex number

$$\langle \gamma, \omega \rangle = \int_\gamma \omega$$

to a piecewise differentiable path γ in a domain D and a differential form ω on D (when the integral exists).

LEMMA 4.12. *Let ω be a differential form on a domain D. Then ω is exact on D if and only if $\int_\gamma \omega = 0$ for all closed and piecewise differentiable paths γ in D.*

PROOF. Assume that ω is exact. Then there exists a C^1-function F on D with

$$\omega = F_x \, dx + F_y \, dy.$$

Let γ be a pdp parameterized by $[a, b]$ joining P_1 to P_2 in D. Then

$$\int_\gamma \omega = \int_a^b \left(F_x \frac{dx}{dt} + F_y \frac{dy}{dt} \right) dt = \int_a^b \frac{dF}{dt} \, dt = F(P_2) - F(P_1),$$

which equals 0 if $P_1 = P_2$.

To prove the converse, let $Z_0 = (x_0, y_0)$ be a fixed point in D and let $Z = (x, y)$ be an arbitrary point in D. Let γ be a pdp in D joining Z_0 to Z. Define

$$F(x, y) = \int_\gamma \omega.$$

Our hypothesis tells us that the function F is well defined on D.

Assume that $\omega = P \, dx + Q \, dy$. We must show that F is C^1 and $dF = \omega$. Choose $\epsilon > 0$ so that $U = U_{(x,y)}(\epsilon) \subset D$; also choose $x_1 \neq x$ such that $(x_1, y) \in U$ and that the straight line L from (x_1, y) to (x, y)

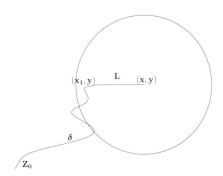

FIGURE 4.2. The integration path for F

is contained in D. Let δ be any pdp in D from (x_0, y_0) to (x_1, y) (see Figure 4.2).

Then

$$F(x, y) = \int_\delta \omega + \int_{x_1}^x P(\xi, y)\, d\xi.$$

It is now clear that $F_x = P$. Similarly, $F_y = Q$.   □

THEOREM 4.13. *Let ω be a differential form on a disk D. Then ω is exact on D if and only if $\int_\gamma \omega = 0$ for all γ that are boundaries of rectangles with sides parallel to the coordinate axes.*

PROOF. Repeat the above argument with (x_0, y_0) the center of D.   □

COROLLARY 4.14. *If D is a disk, then ω is a closed form if and only if it is exact.*

DEFINITION 4.15. A region $R \subseteq \mathbb{R}^2$ is called *(xy)-simple* if it is bounded by a pdp and has the property that any horizontal or vertical line that has nonempty intersection with R intersects it in an interval. Furthermore, the set of values $a \in \mathbb{R}$ for which the line $x = a$ has nonempty intersection with R is an interval, and the set of values $c \in \mathbb{R}$ for which the line $y = c$ has nonempty intersection with R is also an interval. Here an interval may consist of a single point.

In particular, there exist real numbers $c < d$ and functions h_1 and h_2 defined on the interval $[c, d]$ such that the region R may be described as follows:

$$R = \{(x, y); c \le y \le d, h_1(y) \le x \le h_2(y)\}.$$

A similar description may be given interchanging the roles of x and y.

We recall Green's theorem, which will help us to distinguish closed differentials from exact differentials.

THEOREM 4.16 (**Green's theorem**). *Let R be an (xy)-simple region, and let γ denote its boundary oriented counterclockwise. Let $\omega = P\,dx + Q\,dy$ be a C^1-form on a region $D \supset R \cup \gamma$. Then*

$$\iint_R \left(\frac{\partial Q}{\partial x} - \frac{\partial P}{\partial y} \right) dx\,dy = \int_\gamma P\,dx + Q\,dy.$$

PROOF. We have, using the notation introduced in the definition of (xy)-simple regions,

$$\iint_R \frac{\partial Q}{\partial x}\,dx\,dy = \int_c^d \int_{h_1(y)}^{h_2(y)} \frac{\partial Q}{\partial x}\,dx\,dy$$

$$= \int_c^d [Q(h_2(y), y) - Q(h_1(y), y)]\,dy$$

$$= \int_c^d Q(h_2(y), y)\,dy + \int_d^c Q(h_1(y), y)\,dy = \int_\gamma Q\,dy\,.$$

Similarly,

$$\iint_R -\frac{\partial P}{\partial y}\,dx\,dy = -\iint_R \frac{\partial P}{\partial y}\,dy\,dx = \int_\gamma P\,dx\,.$$

□

REMARK 4.17. (1) The theorem can be extended to finite unions of (xy)-simple regions (by cancelation of integrals over common boundaries oppositely oriented).

(2) In terms of complex derivatives, the theorem can be restated as

$$\iint_R \left(\frac{\partial Q}{\partial z} - \frac{\partial P}{\partial \bar{z}} \right) dz\,d\bar{z} = \int_\gamma P\,dz + Q\,d\bar{z}\,,$$

where $dz\,d\bar{z} = -2\imath dx\,dy$.

THEOREM 4.18. *Suppose that $\omega = P\,dx + Q\,dy$ is a C^1-differential form on a domain D. If ω is exact, then $\dfrac{\partial P}{\partial y} = \dfrac{\partial Q}{\partial x}$.* **Conversely,** *if D is an open disk, P and Q are C^1-functions on D, and $\dfrac{\partial P}{\partial y} = \dfrac{\partial Q}{\partial x}$, then $\omega = P\,dx + Q\,dy$ is exact.*

PROOF. Since ω is exact, $\omega = dF = F_x\,dx + F_y\,dy$. But ω is C^1, and thus F is C^2; therefore, $F_{xy} = F_{yx}$.

For the converse, we need only show that $\int_\gamma \omega = 0$ for all closed paths γ in D that are boundaries of rectangles R with sides parallel to the coordinate axes. But

$$\int_\gamma \omega = \iint_R \left(\frac{\partial Q}{\partial x} - \frac{\partial P}{\partial y} \right) dx \, dy = 0.$$

$\square$

REMARK 4.19. Recall that

$$f(z) \, dz = (u + \imath v)(dx + \imath \, dy)$$
$$= (u \, dx - v \, dy) + \imath \, (u \, dy + v \, dx)$$
$$= \omega_1 + \imath \, \omega_2.$$

Thus

$$\int_\gamma f(z) \, dz = \int_\gamma \omega_1 + \imath \int_\gamma \omega_2.$$

Furthermore, $f(z) \, dz$ is closed (exact) if and only if both ω_1 and ω_2 are, and F_j is a primitive for ω_j $(j = 1, 2)$ if and only if $F_1 + \imath F_2$ is a primitive for $f(z) \, dz$.

LEMMA 4.20. *Let $f(z) \, dz$ be of class C^1. Then $f(z) \, dz$ is closed if and only if f is holomorphic.*

PROOF. By the above remarks, $f(z) \, dz$ is closed if and only if $u_y = -v_x$ and $v_y = u_x$ if and only if u and v satisfy CR if and only if f is holomorphic. $\square$

LEMMA 4.21. *A C^1-function F is a primitive for $f(z) \, dz$ if and only if $F' = f$.*

PROOF. The function F is a primitive for $f(z) \, dz$ if and only if $dF = F_z \, dz + F_{\bar{z}} \, d\bar{z} = f(z) \, dz$ if and only if $F_{\bar{z}} = 0$ and $F_z = F' = f$. $\square$

We have now proven the following result.

THEOREM 4.22. *The differential form $f(z) \, dz$ on a domain D is closed if and only if $\int_\gamma f(z) \, dz = 0$, for all boundaries γ of rectangles[3] R contained in D with sides parallel to the coordinate axes.*
If f is C^1, then $f(z) \, dz$ is closed if and only if f is holomorphic.

REMARK 4.23. We shall see that the C^1 assumption is not needed.

EXAMPLE 4.24. Not every closed form is exact. Let $D = \mathbb{C}_{\neq 0}$ and $\omega = \frac{dz}{z}$.

---

[3]It should be emphasized that the rectangle R, not just its perimeter ∂R, is contained in D.

(a) If $\gamma(t) = e^{2\pi i t}$ for $t \in [0,1]$, then $\int_\gamma \omega = 2\pi i$. Thus ω is not exact.

(b) Since $\frac{1}{z}$ is holomorphic and C^1 on D, ω is closed.

Locally $\omega = dF$, where F is a branch of the logarithm. We have produced two real forms, the real and imaginary parts of ω:

$$\frac{dz}{z} = \frac{x\,dx + y\,dy}{x^2 + y^2} + i\,\frac{-y\,dx + x\,dy}{x^2 + y^2} = d\log|z| + i\,d\arg z = d\log z.$$

The first of the two real forms is exact, and the second is closed but not exact on D. Note that $d\arg z = d\arctan\frac{y}{x}$ (for $x \neq 0$). Note also that $\arg z$ and $\arctan\frac{y}{x}$ are multi-valued functions whose differentials agree and are single-valued.

DEFINITION 4.25. Let D be a domain in $\mathbb{C}$, $\gamma : [a,b] \to D$ a continuous path in D and $\omega = P\,dx + Q\,dy$ a closed form in D. A *primitive for ω along γ* is a continuous function $f : [a,b] \to \mathbb{C}$ such that for all $t_0 \in [a,b]$ there exist a neighborhood N of $\gamma(t_0)$ in D and a primitive F for ω in N such that $F(\gamma(t)) = f(t)$ for all t in a neighborhood of t_0 in $[a,b]$.

Caution. It is possible to have $t_1 \neq t_2$ with $\gamma(t_1) = \gamma(t_2)$ but $f(t_1) \neq f(t_2)$; that is, f need not be well defined on γ.

THEOREM 4.26. *If γ is a continuous path in a domain D and ω is a closed form on D, then there exists a primitive f of ω along γ; f is unique up to the addition of a constant.*

PROOF. Suppose γ is parameterized by $[a,b]$ and $\omega = P\,dx + Q\,dy$.

Uniqueness: Suppose f and g are two primitives of ω along γ and let $t_0 \in [a,b]$. Then there exist primitives F and G of ω in a connected neighborhood U of $\gamma(t_0)$ in D such that $F(\gamma(t)) = f(t)$ and $G(\gamma(t)) = g(t)$ for t near t_0.

Hence $F_x = G_x = P$ and $F_y = G_y = Q$ in U, thus $F - G$ restricts to a constant in U and therefore $f - g$ is constant near t_0.

We conclude that $f - g$ is a continuous and locally constant function on the connected set $[a,b]$. Thus $f - g$ is a constant function.

Existence: Given $t \in [a,b]$, there exist an interval $I(t) \subset [a,b]$ (open in $[a,b]$ and containing t), and an open set $U(\gamma(t)) \subset D$ such that ω has a primitive in $U(\gamma(t))$ and $\gamma(I(t)) \subset U(\gamma(t))$. Then

$$\bigcup_{t\in[a,b]} I(t)$$

is an open cover of $[a,b]$, and thus, there exists a finite subcover

$$I_0 \cup I_1 \cup \ldots \cup I_n = [a,b],$$

with corresponding U_j.

Without loss of generality we may assume $I_0 = [a_0, b_0)$, $I_j = (a_j, b_j)$ for $j = 1, \ldots, n-1$, and $I_n = (a_n, b_n]$, where $a_0 = a$, $a_j < b_j$ for $j = 0, \ldots, n$, $b_j > a_{j+1}$ for $j = 0, \ldots, n-1$, and $b_n = b$. Furthermore, $\gamma(I_j) \subset U_j$ and ω has a primitive F_j on U_j for $j = 0, 1, \ldots, n$.

Set $f(t) = F_0(\gamma(t))$ for $t \in I_0$. Having defined f on $I_0 \cup I_1 \cup \ldots \cup I_k$ for $0 \leq k < n$, we define f on I_{k+1} as follows. Let F_{k+1} be any primitive in U_{k+1}, and let F_k be the primitive in U_k such that $f(t) = F_k(\gamma(t))$ for $t \in I_k$. Then F_{k+1} and F_k are primitives for ω in $U_{k+1} \cap U_k$. Thus $F_{k+1} - F_k$ is constant on each connected component of $U_{k+1} \cap U_k$; in particular, $F_{k+1} - F_k = c$ on the component containing $\gamma(I_{k+1} \cap I_k)$. Set $f(t) = F_{k+1}(\gamma(t)) - c$ for $t \in I_{k+1}$; then f is well defined on $I_{k+1} \cap I_k$. $\square$

4.3. Integration of closed forms and the winding number

Consideration of the next example leads to the extension of the integral to more general paths which leads in turn to the surprising result, given in Corollary 4.31, that certain integrals take on only integer values. This fact allows a precise definition corresponding to the intuitive idea of counting the number of times a curve winds around a point, the *winding number* of the curve with respect to a point.

EXAMPLE 4.27. We compute $\int_\gamma \omega$, where ω is closed in D and γ is a pdp in D. As before $\gamma : [a, b] \to D$. Subdivide $I = I_0 \cup I_1 \cup \ldots \cup I_n$, where $I_j = [a_j, a_{j+1}]$, $a_0 = a$ and $a_{n+1} = b$, such that $\gamma_j = \gamma|_{I_j}$ is a differentiable path and ω has a primitive F_j in a neighborhood of $\gamma(I_j)$ for $j = 0, \ldots, n$. Let f be a primitive of ω along γ. Then

$$\int_\gamma \omega = \sum_{j=0}^n \int_{\gamma_j} dF_j = f(b) - f(a).$$

Using the last equation, we extend the concept of a line integral.

DEFINITION 4.28. Let ω be a closed differential form in D and $\gamma : [a, b] \to D$ be a continuous path in D. We define

$$\int_\gamma \omega = f(b) - f(a),$$

where f is a primitive of ω along γ.

REMARK 4.29. The integral is well defined and agrees with the earlier definition for pdps. Note that we have avoided any discussion of rectifiability of the curve γ. We have however paid a price: We have not introduced the class of curves γ whose length is well defined.

THEOREM 4.30. *If γ is any continuous closed path in $\mathbb{C}_{\neq 0}$, then*

$$\frac{1}{2\pi i} \int_\gamma \frac{dz}{z} \in \mathbb{Z}.$$

PROOF. Let f be a primitive of $\dfrac{dz}{z}$ along γ. Then

$$\int_\gamma \frac{dz}{z} = f(b) - f(a),$$

where $[a, b]$ parameterizes γ. Since $\gamma(a) = \gamma(b)$, this difference is just the difference between two branches of $\log z$, hence of the form $2\pi i\, n$ with $n \in \mathbb{Z}$. $\qquad\square$

From Example 4.24, we obtain

COROLLARY 4.31. *For γ as in Theorem 4.30,* $\dfrac{1}{2\pi} \displaystyle\int_\gamma \dfrac{x\,dy - y\,dx}{x^2 + y^2}$ *is an integer.*

DEFINITION 4.32. Let $\zeta \in \mathbb{C}$, and let γ be a continuous closed path in $\mathbb{C} - \{\zeta\}$. We define the *index* or *winding number of γ with respect to ζ* by

$$I(\gamma, \zeta) = \frac{1}{2\pi i} \int_\gamma \frac{dz}{z - \zeta} \in \mathbb{Z}.$$

EXAMPLE 4.33. (In polar coordinates). Let $r = f(\theta) > 0$, with $f \in \mathbf{C}^1(\mathbb{R})$. Let $n \in \mathbb{Z}_{>0}$, and define $\gamma(\theta) = f(\theta)e^{i\theta}$, where $\theta \in [0, 2\pi n]$. Assume that $f(0) = f(2\pi n)$.
Then

$$I(\gamma, 0) = \frac{1}{2\pi i} \int_\gamma \frac{dz}{z} = \frac{1}{2\pi i} \int_0^{2\pi n} \frac{d(f(\theta)e^{i\theta})}{f(\theta)e^{i\theta}}$$

$$= \frac{1}{2\pi i} \int_0^{2\pi n} \frac{f'(\theta)e^{i\theta} + i f(\theta)e^{i\theta}}{f(\theta)e^{i\theta}}\, d\theta = \frac{1}{2\pi i} \int_0^{2\pi n} \left[\frac{f'(\theta)}{f(\theta)} + i \right] d\theta = n\,.$$

In general, let $\gamma : [a, b] \to \mathbb{C} - \{\zeta\}$ be a continuous closed path and let f be a primitive of $\frac{dz}{z - \zeta}$ on γ. Then $f(t)$ agrees with a branch of $\log(\gamma(t) - \zeta)$; that is,

$$e^{f(t)} = \gamma(t) - \zeta \quad \text{for all } t \in [a, b]\,.$$

Hence

$$I(\gamma, \zeta) = \frac{f(b) - f(a)}{2\pi i}\,.$$

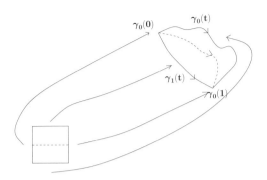

FIGURE 4.3. Homotopy with fixed end points

4.4. Homotopy and simple connectivity

To give the integration results the clearest formulation (see Corollary 4.44), we introduce the concepts of homotopic curves and simply connected domains.

DEFINITION 4.34. Let γ_0 and γ_1 be two continuous paths, in a domain D, parameterized by $I = [0, 1]$ with the same end points; that is, $\gamma_0(0) = \gamma_1(0)$ and $\gamma_0(1) = \gamma_1(1)$. We say that γ_0 and γ_1 are *homotopic (with fixed end points)* if there exists a continuous function $\delta : I \times I \to D$ such that

(1) $\delta(t, 0) = \gamma_0(t)$ for all $t \in I$,
(2) $\delta(t, 1) = \gamma_1(t)$ for all $t \in I$,
(3) $\delta(0, u) = \gamma_0(0) = \gamma_1(0)$ for all $u \in I$, and
(4) $\delta(1, u) = \gamma_0(1) = \gamma_1(1)$ for all $u \in I$.

We call δ a *homotopy with fixed end points* between γ_0 and γ_1; see Figure 4.3.

Let γ_0 and γ_1 be two continuous closed paths in a domain D parameterized by $I = [0, 1]$; that is, $\gamma_0(0) = \gamma_0(1)$ and $\gamma_1(0) = \gamma_1(1)$. We say that γ_0 and γ_1 are *homotopic as closed paths* (see Figure 4.4) if there exists a continuous function $\delta : I \times I \to D$ such that

(1) $\delta(t, 0) = \gamma_0(t)$ for all $t \in I$,
(2) $\delta(t, 1) = \gamma_1(t)$ for all $t \in I$, and
(3) $\delta(0, u) = \delta(1, u)$ for all $u \in I$.

The map δ is called a *homotopy of closed curves or paths.*

A continuous closed path is *homotopic to a point* if it is homotopic to a constant path (as a closed path).

DEFINITION 4.35. Let $I = [0, 1]$, let $\delta : I \times I \to D \subset \mathbb{C}$ be a continuous map, and let ω be a closed form on D.

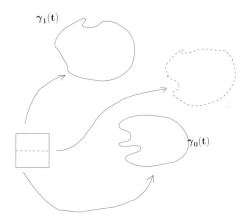

$$\gamma_1(t)$$

$$\gamma_0(t)$$

FIGURE 4.4. Homotopy of closed paths

A function $f : I \times I \to \mathbb{C}$ is said to be a *primitive* for ω along δ provided for every $(t_0, u_0) \in I \times I$, there exists a neighborhood V of $\delta(t_0, u_0)$ in D and a primitive F for ω on V such that $f(t, u) = F(\delta(t, u))$ for all (t, u) in some neighborhood of (t_0, u_0) in $I \times I$.

REMARK 4.36. (1) Such a function f is automatically continuous on $I \times I$.

(2) For fixed $u \in I$, $f(\cdot, u)$ is a primitive for ω along $t \mapsto \delta(t, u)$.

THEOREM 4.37. *If ω is a closed form on D and $\delta : [0,1] \times [0,1] \to D \subset \mathbb{C}$ is a continuous map, then a primitive f for ω along δ exists and is unique up to an additive constant.*

PROOF. We leave the proof as an exercise for the reader.   □

We now observe that all integrals of a closed form along homotopic paths are equal.

THEOREM 4.38. *Let γ_0 and γ_1 be continuous paths in a domain D, and let ω be a closed form on D.*

If γ_0 is homotopic to γ_1 with fixed end points, then

$$\int_{\gamma_0} \omega = \int_{\gamma_1} \omega.$$

PROOF. We assume that both paths are parameterized by $I = [0,1]$. Let $\delta : I \times I \to D$ be a homotopy between our two paths, and let f be a primitive of ω along δ. Thus $u \mapsto f(0, u)$ is a primitive of ω along $u \mapsto \delta(0, u) = \gamma_0(0)$, and hence, $f(0, u)$ is a constant α independent of u. Similarly $f(1, u) = \beta \in \mathbb{C}$.

Now,

$$\int_{\gamma_0} \omega = f(1,0) - f(0,0) = \beta - \alpha$$

and

$$\int_{\gamma_1} \omega = f(1,1) - f(0,1) = \beta - \alpha.$$

$\square$

REMARK 4.39. A similar result holds for two curves that are homotopic as closed paths (see Exercise 4.8).

COROLLARY 4.40. *If γ is homotopic to a point in D and ω is a closed form in D, then*

$$\int_{\gamma} \omega = 0.$$

This corollary motivates the following definition.

DEFINITION 4.41. A region $D \subseteq \mathbb{C}$ is called *simply connected* if every closed path in D is homotopic to a point in D.

EXAMPLE 4.42. (1) The complex plane $\mathbb{C}$ is simply connected. More generally,

(2) disks are simply connected: Let $\zeta \in \mathbb{C}$, $R \in (0, +\infty)$ and $D = \{z \in \mathbb{C}; |z - \zeta| < R\}$. Without loss of generality, $\zeta = 0$ and $R = 1$. Let γ be a closed path in D parameterized by $I = [0,1]$, and define $\delta(t, u) = u\gamma(t)$.

COROLLARY 4.43. *If D is a simply connected domain and γ is a continuous closed path in D, then $\int_{\gamma} \omega = 0$ for all closed forms ω on D.*

We obtain the simplest formulation of the main result:

COROLLARY 4.44. *In a simply connected domain, a differential form is closed if and only if it is exact.*

An immediate corollary gives the existence of branches of the logarithm:

COROLLARY 4.45. *In every simply connected domain not containing the point 0, there exists a branch of $\log z$.*

PROOF. The differential form $\omega = \dfrac{dz}{z}$ is closed and thus exact in the given domain. Hence there exists a holomorphic function F (on the same domain) such that $dF = \omega$. This function F is a branch of the logarithm. $\square$

REMARK 4.46. Annuli and punctured disks are not simply connected. To see this let $R_1, R,$ and R_2 be real numbers with $0 < R_1 < R < R_2$. For any complex number z_0, let A be the annulus $\{z; R_1 < |z - z_0| < R_2\}$ and let D be the punctured disk $\{z; 0 < |z - z_0| < R_2\}$. Then the (range of the) closed path $\gamma(t) = R\exp(2\pi \imath t)$, for $0 \le t \le 1$, is contained in A and in D, and $\int_\gamma \dfrac{dz}{z - z_0} = 2\pi\imath \ne 0$.

To clarify further the concept of simple connectivity, we need

DEFINITION 4.47. A region D is *convex* if every pair of points in D can be joined by a segment in D.

Note that convex implies simply connected, but the converse is not true.

4.5. Winding number

In Section 4.3 we defined the winding number $I(\gamma, \zeta)$ of a curve γ with respect to a point ζ. In this section we will see that it allows us to draw strong conclusions about the behavior of a function defined in a disk (Theorem 4.48).

We begin with some **properties** of $I(\gamma, \zeta)$ for γ a closed path and $\zeta \notin \text{range}\,\gamma$.

(1) If γ_0 and γ_1 are homotopic as closed paths in $D = \mathbb{C} - \{\zeta\}$, then $I(\gamma_0, \zeta) = I(\gamma_1, \zeta)$.

  PROOF. The differential $\omega = \dfrac{dz}{z - \zeta}$ is closed on D. Thus,

  $$\int_{\gamma_0} \omega = \int_{\gamma_1} \omega. \qquad \square$$

(2) If $\gamma : [a, b] \to \mathbb{C}$ is a closed path, then $z \mapsto I(\gamma, z)$ is a locally constant function on $D = \mathbb{C} - \text{range}\,\gamma$; hence, it is constant on each connected component of D.

  PROOF. Let $\zeta \in D$; we need to show that for all $h \in \mathbb{C}$ with $|h|$ sufficiently small, $I(\gamma, \zeta + h) = I(\gamma, \zeta)$.

  Since $\text{range}\,\gamma \subset \mathbb{C} - \{\zeta\}$, for each $t \in [a, b]$, we choose $\delta > 0$ such that $U(\gamma(t), \delta) \subset \mathbb{C} - \{\zeta\}$ and then choose a finite subcover of $\text{range}\,\gamma$.

  Let δ_0 be the Lebesgue number of this cover; that is,

  $$U(\gamma(t), \delta_0) \subset \mathbb{C} - \{\zeta\}$$

  for all $t \in [a, b]$.

If we fix any $h \in \mathbb{C}$ with $|h| < \delta_0$, then

$$\int_\gamma \frac{dz}{z - (\zeta + h)} = \int_\gamma \frac{dz}{(z - h) - \zeta} = \int_{\gamma'} \frac{dz'}{z' - \zeta},$$

where $z' = z - h$ and $\gamma' = \gamma - h$; that is, $\gamma'(t) = \gamma(t) - h$ for all $t \in [a, b]$.

Now γ' is a closed path in $\mathbb{C} - \{\zeta\}$ and γ' is homotopic to γ (as closed paths), via the homotopy defined on $[a, b] \times [0, 1]$ by

$$\delta(t, u) = \gamma(t) - uh\ , t \in [a, b],\ u \in [0, 1].$$

Thus, $I(\gamma, \zeta + h) = I(\gamma', \zeta) = I(\gamma, \zeta)$. $\qquad \square$

(3) If range $\gamma \subset D \subset \mathbb{C} - \{\zeta\}$ with D simply connected, then $I(\gamma, \zeta) = 0$.

PROOF. The differential $\dfrac{dz}{z - \zeta}$ is closed in D, therefore exact. $\qquad \square$

(4) Let $\gamma : \theta \mapsto Re^{i\theta}$, with $R > 0$ and $\theta \in [0, 2\pi]$. Then $I(\gamma, 0) = 1$ [by Example 4.33], $I(\gamma, z) = 1$ for $|z| < R$ [by (2)], and $I(\gamma, z) = 0$ for $|z| > R$ [by (3)].

We now show that if the image of the boundary of a disk under a continuous function f winds non trivially around a point ζ, then f assumes the value ζ someplace inside the disk. More precisely,

THEOREM 4.48. *Let $f : \{z \in \mathbb{C} : |z| \le R\} \to \mathbb{C}$ be a continuous map (with $R > 0$), and let $\gamma(\theta) = f(Re^{2\pi i\theta})$ for $\theta \in [0, 1]$.*

If $\zeta \notin$ range γ and $I(\gamma, \zeta) \ne 0$, then there exists a z such that $|z| < R$ and $f(z) = \zeta$.

PROOF. Assume $f(z) \ne \zeta$ for all $|z| < R$. Then $f(z) \ne \zeta$ for all $|z| \le R$.

Define $\delta(\rho, \theta) = f(\rho Re^{2\pi i\theta})$ on $[0, 1] \times [0, 1]$. Then δ is continuous, $\delta(1, \theta) = \gamma(\theta)$, $\delta(0, \theta) = f(0)$, $\delta(\rho, 0) = \delta(\rho, 1)$ and $\delta(\rho, \theta) \in \mathbb{C} - \{\zeta\}$. Thus γ is homotopic to a point in $\mathbb{C} - \{\zeta\}$, and hence, $I(\gamma, \zeta) = 0$. We have arrived at the needed contradiction. $\qquad \square$

DEFINITION 4.49. Let γ_1 and γ_2 be continuous paths parameterized by $[0, 1]$. We define two new paths, also parameterized by $[0, 1]$:

$$\gamma_1\gamma_2 : t \mapsto \gamma_1(t)\gamma_2(t),$$
$$\gamma_1 + \gamma_2 : t \mapsto \gamma_1(t) + \gamma_2(t).$$

Note that the above definition of a product of two paths differs from the one used in topology, where the paths are traversed in succession but at twice the speed.[4]

THEOREM 4.50. *If γ_1 and γ_2 are continuous closed paths not passing through 0, then*

$$I(\gamma_1\gamma_2, 0) = I(\gamma_1, 0) + I(\gamma_2, 0).$$

PROOF. Let $\omega = \dfrac{dz}{z}$ and $\gamma_j : [0, 1] \to \mathbb{C} - \{0\}$.

Choose continuous functions $f_j : [0, 1] \to \mathbb{C}$ so that $e^{f_j(t)} = \gamma_j(t)$ for all $t \in I$. Then $e^{f_1(t) + f_2(t)} = \gamma_1\gamma_2(t)$. Thus, $f = f_1 + f_2$ is a primitive of ω along $\gamma_1\gamma_2$, and

$$I(\gamma_1\gamma_2, 0) = \frac{f(1) - f(0)}{2\pi\imath} = \frac{f_1(1) - f_1(0)}{2\pi\imath} + \frac{f_2(1) - f_2(0)}{2\pi\imath}.$$

□

THEOREM 4.51. *Let γ_1 and γ be continuous closed paths in $\mathbb{C}$. Assume that*

$$0 < |\gamma_1(t)| < |\gamma(t)| \quad \text{for all } t \in [0, 1].$$

Then

$$I(\gamma_1 + \gamma, 0) = I(\gamma, 0).$$

PROOF. Note that

$$\gamma(t) + \gamma_1(t) = \gamma(t)\left(1 + \frac{\gamma_1(t)}{\gamma(t)}\right) = \gamma(t)\beta(t) \quad \text{for all } t \in [0, 1]$$

with $\beta(t) = 1 + \dfrac{\gamma_1(t)}{\gamma(t)}$.

Now

$$|\beta(t) - 1| < 1,$$

and thus β is a closed path in the simply connected domain $U(1, 1)$. Therefore, $I(\beta, 0) = 0$.

□

---

[4]In topology, if γ_1 and γ_2 are continuous paths parameterized by $[0, 1]$ and $\gamma_1(1) = \gamma_2(0)$, then

$$\gamma_1\gamma_2(t) = \begin{cases} \gamma_1(2t), & \text{for } 0 \le t \le \frac{1}{2}; \\ \gamma_2(2t - 1), & \text{otherwise.} \end{cases}$$

4.6. Cauchy Theory: initial version

The most important technical result of this chapter is the following

THEOREM 4.52 (**Goursat's theorem**). *If f is holomorphic on a domain D, then $f(z)\,dz$ is a closed differential form on D.*

This theorem has many significant consequences. To prove it we need some preliminaries. Recall that the only issue that needs to be addressed involves the smoothness of the function f; we are not assuming that the function has continuous partial derivatives. For C^1-functions, we already have the result (see Theorem 4.22).

We follow a beautiful classic line of reasoning.

DEFINITION 4.53. Let γ be a pdp parameterized by the unit interval $[0,1]$ in a domain D and let f be a continuous function on D.

For such restricted paths,[5] we define

(1) the integral of f on γ with respect to arc-length

$$\int_\gamma f(z)\,|dz| = \int_0^1 f(\gamma(t))\,|\gamma'(t)|\,dt,$$

(2) the path γ traversed backward

$$\gamma_-(t) = \gamma(1-t), \quad \text{for all } t \in [0,1],$$

and
(3) the length of the curve γ

$$L(\gamma) = \text{ length of } \gamma = \int_\gamma |dz|\,.$$

The following results follow straightforwardly from these definitions.

PROPOSITION 4.54. *Let γ be a pdp parameterized by the unit interval $[0,1]$ in a domain D and let f be a continuous function on D. Then*

$$\int_\gamma f(z)\,|dz| = \int_{\gamma_-} f(z)\,|dz|$$

and

$$\left|\int_\gamma f(z)\,dz\right| \leq \int_\gamma |f(z)|\,|dz| \leq (\sup\{|f(z)| : z \in \text{range}\,\gamma\}) \cdot L(\gamma)\,.$$

---

[5]The middle definition is valid for all paths.

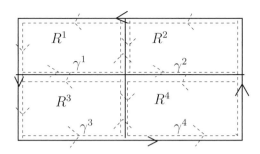

FIGURE 4.5. The integrals along the common side of R^1 and R^2 are in opposite directions

PROOF OF THEOREM 4.52. (Due to E. Goursat). To show that $\omega = f(z)dz$ is a closed form in D, we show that if γ denotes the boundary of R oriented counterclockwise, where R is any rectangle in D with sides parallel to the coordinate axes, then $\int_\gamma \omega = 0$.

Even though the proof of this last statement is due to Goursat, the result is known as *Cauchy's theorem for a rectangle*.

Assume to the contrary that $\int_\gamma \omega = \alpha \neq 0$ and divide R into four congruent rectangles $R^1, \ldots, R^4$, with boundaries $\gamma^1, \ldots, \gamma^4$. Then

$$\alpha = \int_\gamma \omega = \sum_{i=1}^4 \int_{\gamma^i} \omega = \sum_{i=1}^4 \alpha^i.$$

The second equality follows from the fact that certain paths on the boundaries of the subrectangles have opposite directions, giving cancelations in the integrals (see Figure 4.5).

It is clear that for at least one index i, we must have $|\alpha^i| \geq \frac{|\alpha|}{4}$. Call this rectangle R_1, its boundary γ_1, and the corresponding integral α_1. By repeating this procedure, we obtain a sequence of closed rectangles

$$R \supset R_1 \supset \ldots \supset R_k \supset \ldots$$

with boundaries $\gamma_k = \partial R_k$ so that for $\alpha_k = \int_{\gamma_k} \omega$, we have

$$|\alpha_k| \geq \frac{|\alpha|}{4^k}.$$

Each rectangle R_k is a closed subset of D and also of $\mathbb{C}$; furthermore,

$$\lim_{k\to\infty} \text{Area}\,(R_k) = \lim_{k\to\infty} \frac{\text{Area}\,(R)}{4^k} = 0.$$

The Bolzano–Weierstrass Theorem states that there exists a unique $\zeta \in \cap_k R_k$. Since f is holomorphic at ζ,

$$f(z) = f(\zeta) + f'(\zeta)(z - \zeta) + \varepsilon(z)\,|z - \zeta|,$$

where
$$\lim_{z \to \zeta} \varepsilon(z) = 0 \,.$$

Let $\lambda_k = $ length of a diagonal of $R_k = \dfrac{\lambda}{2^k}$, where λ is the length of a diagonal of R. Now

$$\alpha_k = \int_{\gamma_k} f(z)\,dz = f(\zeta)\int_{\gamma_k} dz + f'(\zeta)\int_{\gamma_k}(z-\zeta)\,dz + \int_{\gamma_k}|z-\zeta|\,\varepsilon(z)\,dz.$$

The first two integrals following the second equal sign are zero since the integrands are exact forms (dz and $\dfrac{1}{2}d(z-\zeta)^2$, respectively). Thus, we have

$$\alpha_k = \int_{\gamma_k} |z-\zeta|\,\varepsilon(z)\,dz\,.$$

Now on R_k we have

$$|z - \zeta| \leq \lambda_k = 2^{-k}\,\lambda\,.$$

Given $\epsilon > 0$, there exists a $K \in \mathbb{Z}_{>0}$ such that $|\varepsilon(z)| < \epsilon$ for $z \in R_K$. Thus also for $z \in R_k$ with $k \geq K$. Hence for $k \geq K$, we have

$$|\alpha_k| \leq \int_{\gamma_k} |z-\zeta|\,|\varepsilon(z)|\,|dz| \leq 2^{-k}\,\lambda\,\epsilon\,L(\gamma_k) = \epsilon\,4^{-k}\,\lambda\,L(\gamma)\,.$$

We conclude that

$$|\alpha| \leq 4^k\,|\alpha_k| \leq \epsilon\,\lambda\,L(\gamma).$$

Since ϵ is arbitrary, we must have that $\alpha = 0$.   □

COROLLARY 4.55 (**Cauchy's theorem**). *If f is holomorphic on an open set D and γ is a continuous closed path in D that is homotopic to a point in D, then $\int_{\gamma} f(z)\,dz = 0$.*

COROLLARY 4.56. *If f is holomorphic in a domain D, then locally $f(z)\,dz$ has a primitive in D.*

REMARK 4.57. We have previously shown that (1) implies (4) in the Fundamental Theorem. We have just established another path for obtaining the same conclusion.

Exercises

4.1. Evaluate the line or contour integral $\int_C |z|\,dz$ directly from the definition if

(1) C is a straight line segment from $-\imath$ to $\imath$.
(2) C is the left half of the unit circle traversed from $-\imath$ to $\imath$.
(3) C is the right half of the unit circle traversed from $-\imath$ to $\imath$.

4.2. Evaluate the line or contour integral $\int_C x \, dz$ directly from the definition when C is the line segment from 0 to $i + 1$.

4.3. Evaluate the line or contour integral $\int_C (z - z_0)^m dz$ directly from the definition, where C is the circle centered at z_0 with radius $r > 0$ and

(1) m is an integer, $m \geq 0$.
(2) m is an integer, $m < 0$.

4.4. Evaluate the line or contour integral $\int_\gamma z^3 dz$ directly from the definition over the path $\gamma(t), 0 \leq t \leq 1$, where

(1) $\gamma(t) = 1 + it$.
(2) $\gamma(t) = e^{-\pi it}$.
(3) $\gamma(t) = e^{\pi it}$.
(4) $\gamma(t) = 1 + it + t^2$.

Evaluate the integrals $\int_\gamma \bar{z} \, dz$ and $\int_\gamma \frac{1}{z} dz$ over the same paths.

4.5. Let D_1 and D_2 be simply connected plane domains whose intersection is nonempty and connected. Prove that their intersection and their union are both simply connected.

4.6. Show that for any closed interval $[a, b]$ there exists a one-to-one, onto, and differentiable map $t : [0, 1] \rightarrow [a, b]$.

4.7. Establish Remark 4.7.

4.8. Let D be a domain in $\mathbb{C}$. We have studied the pairing $\int_\gamma \omega$, where γ is a closed path in D and ω is a closed differential form in D. Show that

(1) $\int_\gamma \omega$ depends only on the homotopy class of the closed path γ; that is, if we replace γ by a closed path γ' homotopic to γ (as closed paths), then the integral is unchanged,
(2) (only for those who know some algebraic topology) $\int_\gamma \omega$ depends only on the homology class of the closed path γ; that is, if we replace γ by a closed path γ' homologous to γ, then the integral is unchanged, and
(3) $\int_\gamma \omega$ depends only on the cohomology class of the closed form ω; that is, if we replace ω by $\omega + df$ with $f \in \mathbf{C}^2(D)$, then the integral is unchanged.

4.9.   (1) Let $\gamma : [a, b] \rightarrow \mathbb{C}$ denote a pdp, and let $\varphi : \text{range} \, \gamma \rightarrow \mathbb{C}$ be a continuous function.
          Define $g : D = \mathbb{C} - \text{range} \, \gamma \rightarrow \mathbb{C}$ by

$$g(z) = \int_\gamma \frac{\varphi(u)}{u - z} \, du \, .$$

Show that g has derivatives of all orders and that

$$g^{(n)}(z) = n! \int_\gamma \frac{\varphi(u)}{(u-z)^{n+1}} \, du$$

for all n in $\mathbb{Z}_{\geq 0}$. Thus, in particular, g is holomorphic on D.

(2) Use the above to prove again that if $\gamma(t) = z_0 + R\exp(2\pi it)$ for $0 \leq t \leq 1$, then

$$\frac{1}{2\pi i} \int_\gamma \frac{dz}{z-z_0} = \begin{cases} 1, & \text{if } |z-z_0| < R; \\ 0, & \text{if } |z-z_0| > R. \end{cases}$$

CHAPTER 5

The Cauchy Theory–Key Consequences

This chapter is devoted to some immediate consequences of the Fundamental Theorem for Cauchy Theory, Theorem 4.52, of the last chapter. Although the chapter is very short, it includes proofs of many of the implications of the Fundamental Theorem 1.1. We point out that these relatively compact proofs of a host of major theorems result from the work put in Chapter 4 and earlier chapters.

The appendix to this chapter contains a version of Cauchy's integral formula for smooth (not necessarily holomorphic) functions.

5.1. Consequences of the Cauchy Theory

We begin with a technical strengthening of Theorem 4.52 of the previous chapter allowing functions that are holomorphic on a domain except on a line segment. It will lead to Cauchy's integral formula, once described as the most beautiful theorem in complex variables.

THEOREM 5.1 (**Goursat's theorem, strengthened version**). *If f is continuous in a domain D and holomorphic except on a line segment in D, then $f(z)\,dz$ is closed in D.*

PROOF. Without loss of generality, D is the unit disk and the line segment is all or part of the real axis in D.

We must show that the integral $\int_\gamma f(z)\,dz$ vanishes whenever γ is the boundary of an open rectangle R whose closure is contained in D and whose sides are parallel to the coordinate axes.

There are three possibilities for such rectangles:

(1) The closure of R does not intersect the real axis.
(2) The closure of R has one side on $\mathbb{R}$.
(3) (The interior of) R intersects $\mathbb{R}$.

In case (1) there is nothing to do. In case (3), we reduce to the case of two rectangles as in case (2). Thus it suffices to consider a rectangle of type (2). Assume that the rectangle R lies in the upper semi-disk with one side on $\mathbb{R}$ from a to b. (The possibility of R in the lower half disk is handled similarly.)

83

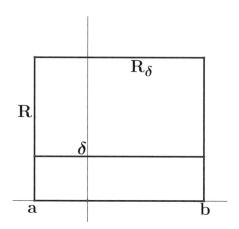

FIGURE 5.1. The rectangles R and R_δ

Let R_δ be the rectangle R with the portion below height δ chopped off, $\delta > 0$, and small (see Figure 5.1). Then the difference of the integrals over the boundary of R and the boundary of R_δ is an integral over a rectangle:

$$
\int_{\partial R} f(z)\, dz - \int_{\partial R_\delta} f(z)\, dz = \int_a^b f(x,0)\, dx + \imath \int_0^\delta f(b,y)\, dy
$$

$$
+ \int_b^a f(x,\delta)\, dx + \imath \int_\delta^0 f(a,y)\, dy
$$

$$
= \int_a^b (f(x,0) - f(x,\delta))\, dx + \imath \int_0^\delta (f(b,y) - f(a,y))\, dy .
$$

Now given $\epsilon > 0$, there exists a δ with $0 < \delta < \epsilon$ such that $|z - \zeta| < \delta$ implies that $|f(z) - f(\zeta)| < \epsilon$ for all z and ζ in R (by the uniform continuity of f on R). Also an $M > 0$ exists such that $|f(z)| \leq M$ for all $z \in R$. Thus

$$
\left| \int_{\partial R} f(z)\, dz - \int_{\partial R_\delta} f(z)\, dz \right| \leq \epsilon(b - a) + 2M\epsilon.
$$

Since ϵ is arbitrary, this tells us that

$$
\int_{\partial R} f(z)\, dz - \int_{\partial R_\delta} f(z)\, dz = 0.
$$

$\square$

We apply this strengthened theorem to obtain

THEOREM 5.2 (**Cauchy's integral formula**). *If f is holomorphic on a domain D and γ is a continuous closed path homotopic to a point in D, then for all $\zeta \in D -$ range γ, we have*

$$\frac{1}{2\pi\imath} \int_\gamma \frac{f(z)}{z - \zeta}\, dz = I(\gamma, \zeta) \cdot f(\zeta). \tag{5.1}$$

PROOF. Define, for $z \in D$,

$$g(z) = \begin{cases} \dfrac{f(z) - f(\zeta)}{z - \zeta}, & \text{if } z \neq \zeta; \\ f'(\zeta), & \text{if } z = \zeta. \end{cases}$$

Then g is continuous on D and holomorphic except (possibly) at ζ, and thus, by Theorem 5.1, $g(z)\, dz$ is closed in D. It follows that

$$0 = \int_\gamma g(z)\, dz = \int_\gamma \frac{f(z) - f(\zeta)}{z - \zeta}\, dz.$$

Thus

$$\int_\gamma \frac{f(z)}{z - \zeta}\, dz = f(\zeta) \int_\gamma \frac{dz}{z - \zeta}.$$

□

EXAMPLE 5.3. Let D be a domain in $\mathbb{C}$, and let f be holomorphic on D. Let $\zeta \in D$. Choose $R > 0$ such that cl $U(\zeta, R) \subset D$, and let $\gamma(\theta) = \zeta + Re^{2\pi\imath\theta}$, $0 \le \theta \le 1$. Then $I(\gamma, w) = 1$ for $|w - \zeta| < R$ and $I(\gamma, w) = 0$ for $|w - \zeta| > R$. Thus

(1)
$$\frac{1}{2\pi\imath} \int_\gamma \frac{f(z)}{z - w}\, dz = f(w) \text{ for } |w - \zeta| < R$$

and

(2)
$$\frac{1}{2\pi\imath} \int_\gamma \frac{f(z)}{z - w}\, dz = 0 \text{ for } |w - \zeta| > R.$$

REMARK 5.4. Equation (5.1) gives the amazing result that the value of a holomorphic function at a point interior to a circle (or eventually any simple closed curve) is determined completely by the values of the function on the boundary circle. The function must, of course, be holomorphic in a neighborhood of the the closed disk bounded by the circle. Note that it then follows from Exercise 4.9 that a holomorphic function has derivatives of all orders. We now prove a more general result.

THEOREM 5.5 (**Power series expansions for holomorphic functions**). *If f is holomorphic in the open disk $\{|z| < R\}$, with R in $(0, +\infty]$, then f has a power series expansion at each point in this disk. In particular, there exists a power series $\sum_{k=0}^{\infty} a_k z^k$ with radius of convergence $\rho \geq R$ such that*

$$f(z) = \sum_{k=0}^{\infty} a_k z^k \quad \text{for } |z| < R.$$

PROOF. It suffices to establish just the particular claim. Choose $0 < r_0 < R$, and define $\gamma(\theta) = r_0 e^{2\pi \imath \theta}$ for $0 \leq \theta \leq 1$. Then $|z| = r < r_0$ implies $I(\gamma, z) = 1$. We start with

$$f(z) = \frac{1}{2\pi \imath} \int_{\gamma} \frac{f(t)}{t - z} \, dt.$$

Now

$$\frac{1}{t-z} - \frac{1}{t} \frac{\left(\frac{z}{t}\right)^{n+1}}{1 - \frac{z}{t}} = \frac{1}{t} \frac{1 - \left(\frac{z}{t}\right)^{n+1}}{1 - \frac{z}{t}} = \frac{1}{t} \sum_{k=0}^{n} \left(\frac{z}{t}\right)^k.$$

Hence

$$\frac{1}{t-z} = \frac{1}{t} \sum_{k=0}^{n} \left(\frac{z}{t}\right)^k + \frac{1}{t} \frac{\left(\frac{z}{t}\right)^{n+1}}{1 - \frac{z}{t}}$$

and

$$f(z) = \frac{1}{2\pi \imath} \left[\sum_{k=0}^{n} z^k \int_{\gamma} \frac{f(t)}{t^{k+1}} \, dt + \int_{\gamma} \frac{f(t)}{t} \frac{\left(\frac{z}{t}\right)^{n+1}}{1 - \frac{z}{t}} \, dt \right]$$

$$= \sum_{k=0}^{n} a_k z^k + R_n,$$

where

$$a_k = \frac{1}{2\pi \imath} \int_{\gamma} \frac{f(t)}{t^{k+1}} \, dt, \quad \text{and} \quad R_n = \frac{1}{2\pi \imath} \int_{\gamma} \frac{f(t)}{t} \frac{\left(\frac{z}{t}\right)^{n+1}}{1 - \frac{z}{t}} \, dt.$$

On γ we have $t = r_0 e^{2\pi \imath \theta}$, $|t| = r_0$ and $|dt| = 2\pi r_0 d\theta$. Let

$$M(r_0) = \sup\{|f(r_0 e^{2\pi \imath \theta})| : 0 \leq \theta \leq 1\},$$

and observe that $\left|1 - \dfrac{z}{t}\right| \geq 1 - \dfrac{r}{r_0}$. Hence

$$|R_n| \leq \frac{1}{2\pi} \int_0^1 \frac{M(r_0)}{r_0} \frac{\left(\frac{r}{r_0}\right)^{n+1}}{1 - \frac{r}{r_0}} r_0 2\pi \, d\theta$$

$$= M(r_0) \frac{\left(\frac{r}{r_0}\right)^{n+1}}{1 - \frac{r}{r_0}} \longrightarrow 0 \quad \text{as} \quad n \to \infty.$$

We conclude that $f(z) = \displaystyle\sum_{k=0}^{\infty} a_k z^k$ for $|z| < R$; furthermore,

$$a_k = \frac{1}{2\pi i} \int_\gamma \frac{f(t)}{t^{k+1}} \, dt$$

is independent of r_0. We have also obtained the estimates

$$|a_k| \leq \frac{M(r_0)}{r_0^k} \tag{5.2}$$

and therefore,

$$|a_k|^{\frac{1}{k}} \leq \frac{M(r_0)^{\frac{1}{k}}}{r_0} .$$

Thus the radius of convergence ρ of $\sum a_k z^k$ satisfies $\rho \geq r_0$ for all $r_0 < R$; in particular, $\rho \geq R$. □

We will make more use of (5.2) shortly.

COROLLARY 5.6. *A function f is holomorphic in an open set D if and only if f has a power series expansion at each point of D. For a holomorphic function f on D, the power series expansion of f at $\zeta \in D$ has radius of convergence*

$$\rho \geq \sup\{r > 0; \ U(\zeta, r) \subset D\}.$$

COROLLARY 5.7. *If f is holomorphic on a domain D, then f is C^∞ in D, and $f^{(n)}$ is holomorphic on D for $n = 0, 1, 2, \ldots$.*

COROLLARY 5.8 (**Cauchy's generalized integral formula**). *Let f be holomorphic on an open set D containing $\operatorname{cl} U(\zeta, R)$ for some $\zeta \in D$ and $R > 0$. If $\gamma(\theta) = \zeta + R e^{i\theta}$ for $0 \leq \theta \leq 2\pi$, then*

$$f^{(n)}(\zeta) = \frac{n!}{2\pi i} \int_\gamma \frac{f(t)}{(t - \zeta)^{n+1}} \, dt$$

for $n = 0, 1, 2, \ldots$.

PROOF. Recall that

$$a_n = \frac{1}{2\pi i} \int_\gamma \frac{f(t)}{(t-\zeta)^{n+1}} \, dt = \frac{f^{(n)}(\zeta)}{n!} \, .$$

$\square$

THEOREM 5.9 (**Morera's theorem**). *If $f \in \mathbf{C}^0(D)$ and $f(z) \, dz$ is closed on D, then f is holomorphic on D.*

PROOF. The differential form $\omega = f(z) \, dz$ is locally exact. Thus for each point $\zeta \in D$, there is a neighborhood U of ζ in D and a primitive F of ω in U. That is, there is a C^1-function F on U with $F_z = f$ and $F_{\bar{z}} = 0$; thus, F is holomorphic on U and so is its derivative f. $\square$

COROLLARY 5.10. *If f is continuous in D and holomorphic except on a line segment in D, then f is holomorphic in D.*

We have by now established the following important

THEOREM 5.11. *Let f be a complex-valued function defined on an open set D in $\mathbb{C}$. Then the following conditions are equivalent:*

(a) f is holomorphic on D.
(b) f is C^1 in D and satisfies CR on D.
(c) f is C^0 in D and $f(z) \, dz$ is closed on D.
(d) f is C^0 in D and f is holomorphic except possibly on a line segment in D.
(e) f has a power series expansion at each point in D.

REMARK 5.12. As a consequence of the theorem, the space $\mathbf{H}(D)$ defined in Chapter 3 (see Definition 3.49) consists precisely of the holomorphic functions on D, and a meromorphic function (an element of $\mathbf{M}(D)$, see Definition 3.51) is locally the ratio of two holomorphic functions.

Recall that we have established the estimates (5.2) from the Cauchy Integral Formula. An immediate corollary is

COROLLARY 5.13 (**Cauchy's inequalities**). *Let*

$$f(z) = \sum_{n=0}^{\infty} a_n (z-\zeta)^n$$

have radius of convergence $\rho > 0$. Then

$$a_n = \frac{f^{(n)}(\zeta)}{n!}$$

and

$$|a_n| = \left| \frac{f^{(n)}(\zeta)}{n!} \right| \le \frac{M(r)}{r^n}, \tag{5.3}$$

for all $0 < r < \rho$, where

$$M(r) = \sup\{|f(z)|; \ |z - \zeta| = r\}.$$

THEOREM 5.14 (**Liouville's theorem**). *A bounded entire function is constant.*

PROOF. Use the Taylor series expansion of the function at the origin and the estimate (5.3). □

THEOREM 5.15 (**Fundamental Theorem of Algebra**). *If P is a polynomial of degree $n \ge 1$, then there exist $a_1, \ldots, a_n \in \mathbb{C}$ and $b \in \mathbb{C}_{\ne 0}$ such that*

$$P(z) = b \prod_{j=1}^{n} (z - a_j) \quad \text{for all} \ \ z \in \mathbb{C}.$$

PROOF. It suffices to show that P has a root. If not, $\dfrac{1}{P}$ is an entire function. It is also bounded since $\lim\limits_{z \to \infty} \dfrac{1}{P(z)} = 0$, and thus, it must be constant. □

5.2. Cycles and homology

In some subsequent chapters we will need a more general form of Cauchy's theorem that deals with integrals over *cycles* that are *homologous* to zero.

DEFINITION 5.16. A *cycle* γ is a finite sequence of continuous oriented closed paths in the complex plane.

We write $\gamma = (\gamma_1, \gamma_2, \ldots, \gamma_n)$ if γ is a cycle and $\gamma_1, \gamma_2, \ldots, \gamma_n$ are the closed paths that make up γ. The paths in a cycle are not necessarily distinct. We will consider the range of γ to be the union of the ranges of the components γ_i.

We extend the notion of the integral of a function over a single closed path to the integral over a cycle as follows:

DEFINITION 5.17. If γ is a cycle and $\gamma_1, \ldots, \gamma_n$ are its components, then for any holomorphic function f defined on a domain U such that range $\gamma \subseteq U$, we set

$$\int_\gamma f(z)\, dz = \int_{\gamma_1} f(z)\, dz + \ldots + \int_{\gamma_n} f(z)\, dz \tag{5.4}$$

We can extend the notion of the index of a point with respect to a path to the index of a point with respect to a cycle by

DEFINITION 5.18. The *index of a point ζ with respect to the cycle* γ is denoted by $I(\gamma, \zeta)$ and defined by

$$I(\gamma, \zeta) = I(\gamma_1, \zeta) + \ldots + I(\gamma_n, \zeta). \tag{5.5}$$

DEFINITION 5.19. A cycle γ with range contained in a domain $U \subseteq \mathbb{C}$ is said to be *homologous to zero with respect to* U if $I(\gamma, \zeta) = 0$ for every $\zeta \in \mathbb{C} - U$.

With these definitions, it is easy to see that Cauchy's theorem can be stated in its most general form as

THEOREM 5.20 (**Cauchy's theorem: general form**). *If f is analytic in a domain $U \subseteq \mathbb{C}$, then $\int_\gamma f(z)dz = 0$ for every cycle γ that is homologous to zero in U.*

REMARK 5.21. A topologist would develop the concept of homology in much more detail using chains and cycles. However, for our purposes, the above definitions suffice. In particular, our cycles are allowed repetitions of the component curves, and the component curves may be taken in any order since we are only concerned with the sum of the integrals.

An additional notion we will use is that of two cycles $\gamma = (\gamma_1, \ldots, \gamma_n)$ and $\delta = (\delta_1, \ldots, \delta_m)$ being homologous in the domain U. Namely, γ and δ (with range contained in U) are *homologous in* U if the cycle with components $(\gamma_1, \ldots, \gamma_n, \delta_{1-}, \ldots, \delta_{m-})$ is homologous to zero in U, where δ_{i-} is the curve δ_i traversed backward (see Definition 4.53).

We also speak of two nonclosed paths γ_1 and γ_2 (with range contained in U) as being *homologous in* U if they have the same initial point and the same end point and the closed cycle (γ_1, γ_{2-}) is homologous to zero with respect to U.

We point out that there is a difference between the cycle $\gamma = (\gamma_1, \gamma_2, \ldots, \gamma_n)$ and the sum $\gamma_1 + \gamma_2 + \cdots \gamma_n$ of the paths in γ, as in Definition 4.49.

5.3. Jordan curves

We introduce some more terminology.

DEFINITION 5.22. Let $\gamma : [0, 1] \to \mathbb{C}$ be a continuous closed path. The curve γ is called a *simple closed path* or a *Jordan* curve whenever $\gamma(t_1) = \gamma(t_2)$ with $0 \le t_1 < t_2 \le 1$ implies $t_1 = 0$ and $t_2 = 1$.

In this case, the range of γ is a homeomorphic image of the unit circle S^1. To see this, we define

$$h(e^{2\pi it}) = \gamma(t),$$

and we note that h maps S^1 onto the range of γ. Observe that h is well defined, continuous, and injective. Since the circle is compact, h is a homeomorphism.

We state, without proof,

THEOREM 5.23 (**Jordan Curve Theorem**[1]). *If γ is a simple closed path in $\mathbb{C}$, then*

(a) $\mathbb{C} - \text{range}\,\gamma$ *has exactly two connected components, one of which is bounded;*

(b) $\text{range}\,\gamma$ *is the boundary of each of these components; and*

(c) $I(\gamma, \zeta) = 0$ *for all ζ in the unbounded component of the complement of the range of γ, and $I(\gamma, \zeta) = \pm 1$ for all ζ in the bounded component of the complement of the range of γ. The choice of sign depends only on the choice of direction for traversal on γ.*

DEFINITION 5.24. For a simple closed path γ in $\mathbb{C}$ we define the *interior of* γ, $i(\gamma)$, to be the bounded component of $\mathbb{C} - \text{range}\,\gamma$, and the *exterior of* γ, $e(\gamma)$, to be the unbounded component of $\mathbb{C} - \text{range}\,\gamma$.

If $I(\gamma, \zeta) = +1$ (respectively, -1) for ζ in $i(\gamma)$, then we say that γ is a *Jordan curve* with *positive (respectively negative) orientation*.

We shall not prove the above theorem. It is a deep result. In all of our applications, it will be obvious that our Jordan curves have the above properties.

REMARK 5.25. If we view a Jordan curve γ as lying on the Riemann sphere $\widehat{\mathbb{C}}$, then each component of the complement of its range is simply connected.

This observation allows us to prove

THEOREM 5.26. [**Cauchy's theorem (extended version)**] *Let $\gamma_0, \ldots, \gamma_n$ be $n+1$ positively oriented Jordan curves. Assume that for all $1 \le j \ne k \le n$,*

$$\text{range}\,\gamma_j \subset e(\gamma_k) \cap i(\gamma_0)$$

(see Figure 5.2).

---

[1]For a proof see the appendix to Ch. IX of J. Dieudonné, *Foundations of Modern Analysis*, Pure and Applied Mathematics, vol. X, Academic Press, 1960, or Chapter 8 of J. R. Munkres, *Topology, a first course*, Prentice-Hall Inc., 1975.

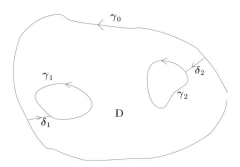

FIGURE 5.2. The Jordan curves and the domain

If f is a holomorphic function on a neighborhood N of the closure of

$$D = i(\gamma_0) \cap e(\gamma_1) \cap \ldots \cap e(\gamma_n),$$

then

$$\int_{\gamma_0} f(z)\,dz = \sum_{k=1}^{n} \int_{\gamma_k} f(z)\,dz.$$

PROOF. Adjoin curves δ_j from γ_0 to γ_j for $j = 1, \ldots, n$. Then the cycle

$$\delta = (\gamma_0, \delta_1, \gamma_{1-}, \delta_{1-}, \ldots, \delta_{n-})$$

is homologous to zero with respect to N. Thus

$$\left(\int_{\gamma_0} + \sum_{j=1}^{n} \left(\int_{\delta_j} + \int_{\gamma_j -} + \int_{\delta_j -} \right) \right) f(z)\,dz = 0 \ .$$

$\square$

An immediate consequence is the following result.

THEOREM 5.27. [**Cauchy's integral formula (extended version)**] *With the hypotheses as in the extended version of Cauchy's theorem 5.26, we have for $\zeta \in D$*

$$2\pi\imath f(\zeta) = \int_{\gamma_0} \frac{f(z)}{z - \zeta}\,dz - \sum_{k=1}^{n} \int_{\gamma_k} \frac{f(z)}{z - \zeta}\,dz \ .$$

PROOF. We can apply Theorem 5.2 to the function f using the neighborhood N of Theorem 5.26 and the cycle δ constructed in its proof, since δ is homologous to zero in N and $I(\delta, \zeta) = +1$. As before an integral over a δ_i is canceled by the corresponding integral over δ_{i-}.

$\square$

5.4. The Mean Value Property

The next concept applies in a broader context than that of holomorphic functions, as we will see in Chapter 9.

DEFINITION 5.28. Let f be a function defined on a domain D in $\mathbb{C}$. We say that f has the *Mean Value Property* (MVP) if for each $\zeta \in D$ there exists $r_0 > 0$ with $U(\zeta, r_0) \subset D$ and

$$f(\zeta) = \frac{1}{2\pi} \int_0^{2\pi} f(\zeta + r\, e^{i\theta})\, d\theta \text{ for all } 0 \le r < r_0. \qquad (5.6)$$

REMARK 5.29. A holomorphic function f on a domain D has the Mean Value Property (with $r_0 = $ the distance of $\zeta \in D$ to ∂D). Hence so do its real and imaginary parts.

THEOREM 5.30 (**Maximum Modulus Principle (MMP)**). *Suppose f is a continuous complex-valued function defined on a domain D in $\mathbb{C}$ that satisfies the Mean Value Property.*

If $|f|$ has a relative maximum at a point $\zeta \in D$, then f is constant in a neighborhood of ζ.

PROOF. The result is clear if $f(\zeta) = 0$. If not, replacing f by $e^{-i\theta} f$ for some $\theta \in \mathbb{R}$, we may assume that $f(\zeta) > 0$.

Write $f = u + iv$, and choose $r_0 > 0$ such that

(1) $\operatorname{cl} U(\zeta, r_0) \subset D$,
(2) (5.6) holds, and
(3) $|f(z)| \le f(\zeta)$ for $z \in \operatorname{cl} U(\zeta, r_0)$.

If we define

$$M(r) = \sup\{|f(z)| \,;\, |z - \zeta| = r\} \text{ for } 0 \le r \le r_0,$$

then

$$M(r) \le f(\zeta) \text{ for } 0 \le r \le r_0.$$

The MVP implies that

$$f(\zeta) = \frac{1}{2\pi} \int_0^{2\pi} f(\zeta + r\, e^{i\theta})\, d\theta \text{ for } 0 \le r < r_0,$$

and thus $f(\zeta) \le M(r)$. We conclude that $f(\zeta) = M(r)$ for $0 \le r \le r_0$.

Now

$$\frac{1}{2\pi} \int_0^{2\pi} M(r)\, d\theta = M(r) = f(\zeta) = \frac{1}{2\pi} \int_0^{2\pi} u(\zeta + r\, e^{i\theta})\, d\theta, \qquad (5.7)$$

where the last equality holds because $f(\zeta)$ is real. Also, it follows from the definition of $M(r)$ that

$$M(r) - u(\zeta + r\, e^{i\theta}) \ge 0. \qquad (5.8)$$

But it also follows from (5.7) that $\int_0^{2\pi} \left[M(r) - u(\zeta + re^{i\theta}) \right] d\theta = 0$ and hence from (5.8) that we must have equality there. Finally,

$$M(r) \geq (u^2(\zeta + r\,e^{i\theta}) + v^2(\zeta + r\,e^{i\theta}))^{\frac{1}{2}} = (M(r)^2 + v^2(\zeta + r\,e^{i\theta}))^{\frac{1}{2}},$$

which implies that $v^2(\zeta + r\,e^{i\theta}) = 0 = v(\zeta + r\,e^{i\theta})$ for $0 \leq r \leq r_0$ and $0 \leq \theta \leq 2\pi$.   □

From the above it is easy to deduce that *a nonconstant holomorphic function on a bounded domain (since it satisfies the MVP) that is continuous on the closure of the domain assumes its maximum on the boundary of that domain.*

COROLLARY 5.31. *Suppose D is a bounded domain and $f \in \mathbf{C}^0(\mathrm{cl}\, D)$ satisfies the MVP in D.*
If

$$M = \sup\{|f(z)| \,;\, z \in \partial D\},$$

then

(a) $|f(z)| \leq M$ *all $z \in D$, and*
(b) *if $|f(\zeta)| = M$ for some $\zeta \in D$, then f is constant in D.*

PROOF. Let

$$M' = \sup\{|f(z)| \,;\, z \in \mathrm{cl}\, D\}\,.$$

Then

$$M \leq M' < +\infty\,.$$

We know that there exists a ζ in $\mathrm{cl}\, D$ such that $|f(\zeta)| = M'$. If $\zeta \in D$, then f is constant in a neighborhood of ζ.
Let

$$D' = \{z \in D;\ |f(z)| = M'\}\,.$$

The set D' is closed and open in D. If nonempty, it is all of D. If $D' = \emptyset$, then $M = M'$. If $D' \neq \emptyset$, then f is constant on D and thus also on $\mathrm{cl}\, D$.   □

In particular, since a function that is holomorphic in a domain satisfies the MVP, we have

COROLLARY 5.32 (**The Maximum Principle for analytic functions**). *If f is a nonconstant holomorphic function on a domain D, then $|f(z)|$ has no maximum in D.*
Furthermore, if D is bounded and f is continuous on the boundary of D, then $|f(z)|$ assumes its maximum on the boundary of D.

REMARK 5.33. By studying the proof of Theorem 5.30, one can prove the *Maximum Principle*, which is an interesting result that may be stated as follows.

Suppose f is a continuous, real-valued function defined on a domain D in $\mathbb{C}$ that satisfies the Mean Value Property. If f has a relative maximum at a point $\zeta \in D$, then f is constant in a neighborhood of ζ.

Similarly, the *Minimum Principle* asserts that a continuous real-valued function defined on a domain D in $\mathbb{C}$ that satisfies the Mean Value Property on D and has a relative minimum at a point $\zeta \in D$ must be constant in a neighborhood of ζ (apply the Maximum Principle to the negative of the function).

An important consequence of Corollary 5.31 is

THEOREM 5.34 (**Schwarz's lemma**). *If f is holomorphic on $\mathbb{D} = U(0,1)$ and satisfies $|f(z)| < 1$ for $|z| < 1$ and $f(0) = 0$, then $|f(z)| \leq |z|$ for $|z| < 1$ and $|f'(0)| \leq 1$.*

Furthermore, if $|f(\zeta)| = |\zeta|$ for some ζ with $0 < |\zeta| < 1$ or if $|f'(0)| = 1$, then there exists a $\lambda \in \mathbb{C}$ with $|\lambda| = 1$ such that $f(z) = \lambda z$ for all $|z| < 1$.

PROOF. Using the Taylor series expansion, we write $f(z) = \sum_{n=1}^{\infty} a_n z^n$; this power series has radius of convergence $\rho \geq 1$. Then

$$g(z) = \begin{cases} \dfrac{f(z)}{z}, & \text{for } z \neq 0; \\ a_1 = f'(0), & \text{for } z = 0, \end{cases}$$

satisfies

$$g(z) = \sum_{n=1}^{\infty} a_n z^{n-1}$$

and is holomorphic on $U(0,1)$.

Now for any r with $0 < r < 1$,

$$|g(z)| = \left| \frac{f(z)}{z} \right| \leq \frac{1}{r} \quad \text{for} \quad |z| = r$$

and thus also for $|z| \leq r$ by MMP. Hence $|g(z)| \leq 1$ or, equivalently, $|f(z)| \leq |z|$ and $|f'(0)| \leq 1$.

If $|g(\zeta)| = 1$ for some ζ with $|\zeta| < 1$, then g is constant, again by MMP. $\square$

REMARK 5.35. In Schwarz's lemma, the hypothesis $|f(z)| < 1$ can be replaced by $|f(z)| \leq 1$. Under this weaker hypothesis, the stronger one still holds, for otherwise there exists an $a \in \mathbb{D}$ with $|f(a)| = 1$ and by MMP, the function f must be constant — obviously impossible.

5.5. On elegance and conciseness

The Jordan Curve Theorem is a major result in two-dimensional topology. All the other theorems and corollaries of this chapter are milestones in function theory. They all were established in an elegant, short, and concise way as a consequence of Goursat's theorem (4.52).

5.6. Appendix: Cauchy's integral formula for smooth functions

In most real analysis courses and books (see, for example, Th. 5.12 of G. B. Folland *Advanced Calculus*, Prentice Hall, 2002), one establishes (we are using complex notation) the following form of

THEOREM 5.36 (**Green's theorem**). *Let D be a compact set in $\mathbb{C}$ that is the closure of its interior, with piecewise smooth positively oriented boundary ∂D (this means that D lies to the left of the oriented curves on its boundary).*
If f and g are C^1-functions on a neighborhood of D, then

$$\iint_D (g_z - f_{\bar{z}}) dz d\bar{z} = \int_{\partial D} f(z) dz + g(z) d\bar{z}.$$

As a consequence of this result, we prove

THEOREM 5.37 (**Cauchy's integral formula for smooth functions**). *Let D be a compact set in $\mathbb{C}$ that is the closure of its interior, with piecewise smooth positively oriented boundary ∂D.*
If f is a C^1-function on a neighborhood of D and ζ is in the interior of D, then

$$f(\zeta) = \frac{1}{2\pi i} \left[\int_{\partial D} \frac{f(z)}{z - \zeta} dz + \iint_D \frac{f_{\bar{z}}(z)}{z - \zeta} dz d\bar{z} \right]. \tag{5.9}$$

PROOF. Choose $\epsilon > 0$ such that the closure of the ball $U(\zeta, \epsilon)$ is contained in the interior of D, and let $D_\epsilon = D - U(\zeta, \epsilon)$.

We apply Green's theorem to the smooth differential form $\dfrac{f(z)}{z - \zeta} dz$ on D_ϵ and obtain

$$\int_{\partial D_\epsilon} \frac{f(z)}{z - \zeta} dz = -\iint_{D_\epsilon} \frac{f_{\bar{z}}(z)}{z - \zeta} dz d\bar{z}. \tag{5.10}$$

Now, ∂D_ϵ consists of two components: ∂D and the clockwise oriented circle with center at ζ and radius ϵ. Hence

$$\int_{\partial D_\epsilon} \frac{f(z)}{z - \zeta}\, dz = \int_{\partial D} \frac{f(z)}{z - \zeta}\, dz - \imath \int_0^{2\pi} f(\zeta + \epsilon e^{\imath\theta})\, d\theta.$$

Letting $\epsilon \to 0$ in (5.10) yields (5.9). $\square$

REMARK 5.38.
- For holomorphic functions, (5.9) reduces to (5.1).
- The last result shows that the Cauchy Theory is a consequence of Green's theorem for C^1-functions that satisfy CR.
- The area element is given by $dz\, d\bar z = -2\imath\, dx\, dy$. This equation can be taken as the definition of the LHS (it really follows from the rules of the exterior differential calculus).

Exercises

5.1. Let D be a domain in $\mathbb{C}$. Prove that the following conditions are equivalent.
(1) D is simply connected.
(2) $\mathbb{C} - D$ is connected.
(3) For each holomorphic function f on D such that $f(z) \neq 0$ for all $z \in D$, there exists a $g \in \mathbf{H}(D)$ such that $f = e^g$ (g is a logarithm of f).
(4) For each holomorphic function f on D such that $f(z) \neq 0$ for all $z \in D$ and for each positive integer n, there exists an $h \in \mathbf{H}(D)$ such that $f = h^n$.

How unique are the functions g and h?

5.2. Let f be analytic in a simply connected domain D, and let γ be a closed piecewise smooth path in D. Set $\beta = f \circ \gamma$. Show that $I(\beta, \zeta) = 0$ for all $\zeta \in \mathbb{C}, \zeta \notin f(D)$.

5.3. Show that if a continuous closed curve γ is homotopic to a point in a domain U, then the cycle (γ) is homologous to zero in U.

Is the converse true?

5.4. Let f be a holomorphic function on $|z| < 1$ with $|f(z)| < 1$ for all $|z| < 1$.
(1) Prove the *invariant form* of Schwarz's lemma (also known as the Schwarz–Pick Lemma):

$$\frac{|f'(z)|}{1 - |f(z)|^2} \leq \frac{1}{1 - |z|^2} \qquad \text{for all } |z| < 1.$$

(2) Find necessary and sufficient conditions for equality in the last equation.

(3) If $f\left(\frac{1}{2}\right) = \frac{1}{3}$, find a sharp upper bound for $\left|f'\left(\frac{1}{2}\right)\right|$.

5.5. Let f be a holomorphic function on $U(0, R)$, $R > 0$. Assume there exist an $M \in \mathbb{R}_{>0}$ such that $|f(z)| \leq M$ for all $z \in U(0, R)$ and an $n \in \mathbb{Z}_{\geq 0}$ such that

$$0 = f(0) = f'(0) = \ldots = f^{(n)}(0).$$

(1) Prove that

$$|f(z)| \leq M \left(\frac{|z|}{R}\right)^{n+1} \quad \text{for all } z \in U(0, R);$$

with equality if and only if there exists an $\alpha \in \mathbb{C}$, $|\alpha| = 1$, such that $f(z) = \alpha M \left(\frac{z}{R}\right)^{n+1}$ for all $z \in U(0, R)$.

(2) Assume that either $f(\zeta) = M \left(\frac{|\zeta|}{R}\right)^{n+1}$ for some ζ with $0 < |\zeta| < R$ or $\left|f^{(n+1)}(0)\right| = (n+1)!M$.
Prove that there exists an $\alpha \in \mathbb{C}$, $|\alpha| = 1$, such that $f(z) = \alpha M \left(\frac{z}{R}\right)^{n+1}$ for all $z \in U(0, R)$.

5.6. Let f be a holomorphic function on the punctured plane $0 < |z| < \infty$. Assume that there exist a positive constant C and a real constant M such that

$$|f(z)| \leq C |z|^M \text{ for } 0 < |z| < \frac{1}{2}.$$

Show that $z = 0$ is either a pole or a removable singularity for f, and find sharp bounds for $\nu_0(f)$, the order of f at 0.

5.7. Prove by use of Schwarz's lemma that every one-to-one conformal mapping of a disk onto another disk is given by a fractional linear transformation. Here the term "disk" is also meant to include half-planes (with ∞ adjoined), and a fractional linear transformation is a map of the form $T(z) = \frac{az+b}{cz+d}$ where $a, b, c,$ and d are complex numbers with $ad - bc \neq 0$.

5.8. Assume that an entire function that takes on every complex value once and only once must be a polynomial of degree one. Prove that the inverse of an entire function cannot be entire except if it is a polynomial of degree one.

5.9. Let f be an entire function with $|f(z)| \leq a|z|^b + c$ for all z, where a, b, and c are positive constants. Prove that f is a polynomial of degree at most b.

5.10. Let f be an entire function such that $f(0) = 0$, and that $\Re(f(z)) \to 0$ as $|z| \to \infty$. Show that f is identically 0.

5.11. Let D be a bounded domain in $\mathbb{C}$. Let $f : \operatorname{cl} D \to \mathbb{C}$ be a nonconstant continuous function, which is analytic in D and satisfies $|f(z)| = 1$ for all $z \in \partial D$. Show that $f(z_0) = 0$ for some $z_0 \in D$.

5.12. Prove the Maximum and the Minimum Principles stated in Remark 5.33.

Furthermore, if D is bounded and f is continuous on the closure of D and satisfies the MVP in D, with $m \leq f \leq M$ on ∂D for some real constants m and M, show that then $m \leq f \leq M$ on D.

CHAPTER 6

Cauchy Theory: Local Behavior and Singularities of Holomorphic Functions

In this chapter we use the Cauchy Theory to study functions that are holomorphic on an annulus and analytic functions with isolated singularities. We describe a classification for isolated singularities. Functions that are holomorphic on an annulus have *Laurent series* expansions, an analog of power series expansions for holomorphic functions on disks. Holomorphic functions with a finite number of isolated singularities in a domain can be integrated using the *Residue Theorem*, an analog of the Cauchy Integral Formula. We discuss the local properties of these functions. The study of zeros of meromorphic functions leads to a theorem of Rouché that connects the number of zeros and poles to an integral. The theorem is not only aesthetically pleasing in its own right but also allows us to give alternative proofs of many important results. In the last section of this chapter we illustrate the use of complex function theory in the evaluation of real definite integrals.

6.1. Functions holomorphic on an annulus

THEOREM 6.1 (**Laurent series expansion**). *Let $\zeta \in \mathbb{C}$, and let f be holomorphic in the annulus*

$$A = \{z \in \mathbb{C};\ 0 \leq R_1 < |z - \zeta| < R_2 \leq +\infty\}.^1$$

Then

$$f(z) = \sum_{n=-\infty}^{\infty} a_n(z - \zeta)^n$$

for all $z \in A$, where the series converges uniformly and absolutely on compact subsets of A, and

$$a_n = \frac{1}{2\pi \imath} \int_{\gamma_r} \frac{f(t)}{(t - \zeta)^{n+1}}\ dt\ ,\quad \text{for } R_1 < r < R_2$$

with

$$\gamma_r = \zeta + re^{i\theta}, 0 \leq \theta \leq 2\pi\ .$$

---

[1]We are including here the cases of *degenerate* annuli: those with $R_1 = 0$ and/or $R_2 = +\infty$.

This series is called a *Laurent* series for f. It is uniquely determined by f and A.

PROOF. Without loss of generality we assume $\zeta = 0$. Consider two concentric circles $\gamma_{r_j} = \{z; |z| = r_j\}$ ($j = 1, 2$), bounding a smaller annulus

$$R_1 < r_1 < |z| < r_2 < R_2.$$

If for $j \in \{1, 2\}$ we let

$$f_j(z) = \frac{1}{2\pi\imath} \int_{\gamma_{r_j}} \frac{f(t)}{t - z} \, dt,$$

then it follows from the extended version of Cauchy's Integral Formula that

$$f(z) = \frac{1}{2\pi\imath} \int_{\gamma_{r_2}} \frac{f(t)}{t - z} \, dt - \frac{1}{2\pi\imath} \int_{\gamma_{r_1}} \frac{f(t)}{t - z} \, dt = f_2(z) - f_1(z).$$

Since f_2 can be extended to be holomorphic in the disk $\{z; |z| < r_2\}$, by Exercise 4.9, we obtain

$$f_2(z) = \sum_{n=0}^{\infty} a_n z^n, \quad \text{with } a_n = \frac{1}{2\pi\imath} \int_{\gamma_{r_2}} \frac{f(t)}{t^{n+1}} \, dt;$$

furthermore, this series converges for $|z| < R_2$ and a_n is independent of r_2 (since any two circles about 0 in A are homotopic in A).

As for f_1, note that

$$-\frac{1}{t - z} = \frac{1}{z - t} = \frac{1}{z} \frac{1}{1 - \frac{t}{z}} = \frac{1}{z} \sum_{k=0}^{\infty} \left(\frac{t}{z}\right)^k \quad \text{for } |t| = r_1 \text{ and } |z| > r_1.$$

Thus $f_1(z) = \frac{1}{2\pi\imath} \frac{1}{z} \sum_{k=0}^{\infty} \int_{\gamma_{r_1}} \frac{f(t)}{z^k} t^k \, dt = \sum_{k=0}^{\infty} z^{-k-1} \frac{1}{2\pi\imath} \int_{\gamma_{r_1}} f(t) t^k \, dt.$

Letting $-k - 1 = n$, we obtain

$$f_1(z) = \sum_{n \leq -1} a_n z^n, \quad \text{where } a_n = \frac{1}{2\pi\imath} \int_{\gamma_{r_1}} \frac{f(t)}{t^{n+1}} \, dt \text{ is independent of } r_1.$$

Observe that f_1 can be extended to be holomorphic in $|z| > R_1$, including ∞, and $f_1(\infty) = 0$. □

COROLLARY 6.2. *If f is holomorphic in A, then $f = f_1 + f_2$ where f_2 is holomorphic in $|z - \zeta| < R_2$ and f_1 is holomorphic in $R_1 < |z - \zeta|$ (including the point at ∞). The functions f_j are unique if we insist that $f_1(\infty) = 0$.*

PROOF. Once again, without loss of generality $\zeta = 0$.

Existence: Already done.

Uniqueness: Suppose $f = f_2 + f_1 = g_2 + g_1$ with appropriate f_i and g_j. Then $0 = (f_2 - g_2) + (f_1 - g_1)$ in A, the function $(f_2 - g_2)$ is holomorphic in $|z| < R_2$, and the function $(f_1 - g_1)$ is holomorphic in $|z| > R_1$ and vanishes at ∞.

Define
$$h(z) = \begin{cases} (g_1 - f_1)(z), & \text{if } |z| > R_1; \\ (f_2 - g_2)(z), & \text{if } |z| < R_2. \end{cases}$$

The function h is well defined and holomorphic on $\mathbb{C} \cup \{\infty\}$ and vanishes at ∞. Hence it is identically zero. $\qquad\qquad\square$

6.2. Isolated singularities

DEFINITION 6.3. We now consider the special case of functions holomorphic on a degenerate annulus with $R_1 = 0$ and $R_2 \in (0, +\infty]$; that is, we fix a point $\zeta \in \mathbb{C}$ and a holomorphic function f on the punctured disk $\{z \in \mathbb{C}; 0 < |z - \zeta| < R_2\}$. In this case, ζ is called an *isolated singularity* of f.

We know that in this case f has the Laurent series expansion

$$f(z) = \sum_{n=-\infty}^{\infty} a_n (z - \zeta)^n \ \text{ for } 0 < |z - \zeta| < R_2 \,.$$

There are three possibilities for the coefficients $\{a_n\}_{n \in \mathbb{Z}_{<0}}$; we now analyze each possibility.

(1) If $a_n = 0$ for all n in $\mathbb{Z}_{<0}$, then f has a *removable singularity* at $z = \zeta$, and f can be extended to be a holomorphic function in the disk $|z - \zeta| < R_2$ by defining $f(\zeta) = a_0$.

Shortly (in Theorem 6.5), we will establish a useful criterion for proving that an isolated singularity is removable.

(2) Finitely many nonzero coefficients with negative indices appear in the Laurent series; that is, there exists N in $\mathbb{Z}_{>0}$ such that $a_{-n} = 0$ for all $n > N$ and $a_{-N} \neq 0$. We can hence write

$$f(z) = \sum_{n=-N}^{-1} a_n (z - \zeta)^n + \sum_{n=0}^{\infty} a_n (z - a)^n$$

for $0 < |z - \zeta| < R_2$.

In this case, $\displaystyle\sum_{n=-N}^{-1} a_n (z - \zeta)^n$ is called the *principal part of* f *at* ζ, f is meromorphic in the disk $|z - \zeta| < R_2$, and f has a pole of order N at $z = \zeta$. Furthermore,

$$\lim_{z \to \zeta} (z - \zeta)^N f(z) = a_{-N} \neq 0$$

and N is characterized by this property (that the limit exists and is different from zero).

(3) Infinitely many nonzero coefficients with negative indices appear in the Laurent series. Then ζ is called an *essential singularity* of f.

EXAMPLE 6.4. We have

$$\exp\left(\frac{1}{z}\right) = \sum_{n=0}^{\infty} \frac{z^{-n}}{n!}$$

for $|z| > 0$. Here $R_1 = 0$ and $R_2 = +\infty$; 0 is an essential singularity of the function, and ∞ is a removable singularity ($f(\infty) = 1$).

THEOREM 6.5. *Assume that*

$$f(z) = \sum_{n=-\infty}^{\infty} a_n z^n \quad \text{for all } 0 < |z| < R_2$$

is the Laurent series of a holomorphic function on a punctured disk. If there exist an $M > 0$ and $0 < r_0 < R_2$ such that

$$|f(z)| \leq M \quad \text{for } 0 < |z| < r_0,$$

then f has a removable singularity at $z = 0$.

PROOF. We know that $a_n = \dfrac{1}{2\pi i} \displaystyle\int_{\gamma_r} \dfrac{f(t)}{t^{n+1}} \, dt$, where $\gamma_r(\theta) = r\,e^{i\theta}$, for $\theta \in [0, 2\pi]$ and $0 < r < r_0$.

We estimate $|a_n| \leq \frac{M}{r^n}$. For $n < 0$, we let $r \to 0$ and conclude that $a_n = 0$. □

THEOREM 6.6 (**Casorati–Weierstrass**). *If f is holomorphic on $\{0 < |z| < R_2\}$ and has an essential singularity at $z = 0$, then for all $c \in \mathbb{C}$ the function*

$$g(z) = \frac{1}{f(z) - c}$$

is unbounded in any punctured neighborhood of $z = 0$.

Therefore the range of f restricted to any such neighborhood is dense in $\mathbb{C}$.

PROOF. Assume that, for some $c \in \mathbb{C}$, the function g is unbounded in some punctured neighborhood N of $z = 0$.

Then there is $\varepsilon > 0$ such that $N = U(0, \varepsilon) - \{0\}$, and, for any $M > 0$, there exists a $z \in N$ such that $|g(z)| > M$; that is, such that $|f(z) - c| < \frac{1}{M}$. Thus c is a limit point of $f(N)$, and the last statement in the theorem is proved.

Now it suffices to prove that for all $c \in \mathbb{C}$ and all $\varepsilon > 0$, the function g is unbounded in $U(0, \varepsilon) - \{0\}$. If g were bounded in such a neighborhood, it would have a removable singularity at $z = 0$ and thus would extend to a holomorphic function on $U(0, \varepsilon)$; therefore f would be meromorphic there.   □

A much stronger result (the next theorem) can be established. This is usually done in more advanced text books.

THEOREM 6.7 (**Picard**). *If f is holomorphic in $0 < |z| < R_2$ and has an essential singularity at $z = 0$, then there exists a $c_0 \in \mathbb{C}$ such that for all $c \in \mathbb{C} - \{c_0\}, f(z) = c$ has infinitely many solutions in $0 < |z| < R_2$.*

EXAMPLE 6.8. The function $\exp\left(\dfrac{1}{z}\right)$ shows the above theorem is sharp ($c_0 = 0$).

DEFINITION 6.9. A function f has an *essential singularity at* ∞ if $g(z) = f\left(\frac{1}{z}\right)$ has an essential singularity at $z = 0$.

EXAMPLE 6.10. For an entire function $f(z) = \displaystyle\sum_{n=0}^{\infty} a_n z^n$ (its radius of convergence ρ equals $+\infty$), there are two possibilities:

(a) Either there exists an N such that $a_n = 0$ for all $n > N$; in which case, f is a polynomial of degree $\leq N$. If $\deg f = N \geq 1$, then f has a pole of order N at ∞. If $\deg f = 0$, f is constant, of course.

(b) Or f has an essential singularity at ∞.

DEFINITION 6.11. If f has an isolated singularity at ζ, with Laurent series $f(z) = \displaystyle\sum_{n=-\infty}^{+\infty} a_n (z - \zeta)^n$ in $0 < |z - \zeta| < R_2$, we define the *residue of f at ζ* by the formula

$$\operatorname{Res}(f, \zeta) = a_{-1} .$$

THEOREM 6.12. *Let A denote the annulus $R_1 < |z - \zeta| < R_2$. If γ is a closed path in A and if f is holomorphic in A with Laurent series*

$$f(z) = \sum_{n=-\infty}^{+\infty} a_n(z - \zeta)^n, \text{ then}$$

$$\frac{1}{2\pi\imath} \int_\gamma f(z) \, dz = I(\gamma, \zeta) \, a_{-1}.$$

In the special case that $R_1 = 0$ and $I(\gamma, \zeta) = 1$, we have

$$\frac{1}{2\pi\imath} \int_\gamma f(z) \, dz = \text{Res}(f, \zeta).$$

PROOF. We write

$$f(z) = \frac{a_{-1}}{z - \zeta} + g(z)$$

where

$$g(z) = \sum_{n \neq -1} a_n(z - \zeta)^n .$$

The function g has a primitive in the annulus; namely,

$$\sum_{n \neq -1} \frac{1}{n+1} a_n (z - \zeta)^n .$$

Thus $\int_\gamma g(z) \, dz = 0.$ ☐

THEOREM 6.13 (**Residue Theorem**). *Let f be holomorphic in a domain D except for isolated singularities at $z_1, \ldots, z_n \in D$. Let γ be a positively oriented Jordan curve homotopic to a point in D such that $z_j \in i(\gamma)$ for $j = 1, \ldots, n$. Then*

$$\int_\gamma f(z) \, dz = 2\pi\imath \sum_{j=1}^n \text{Res}(f, z_j).$$

PROOF. Put a small positively oriented circle around each z_j and use the extended version of Cauchy's theorem. ☐

6.3. Zeros and poles of meromorphic functions

Let D be a domain in $\mathbb{C}$ and $f : D \to \widehat{\mathbb{C}}$ be a meromorphic function. This means that f is holomorphic except for isolated singularities in D, which are removable or poles (see Section 3.5). We have denoted the set (field) of meromorphic functions on D by $\mathbf{M}(D)$.

We now know that at each point of D, f has a Laurent series expansion with only finitely many nonzero coefficients for negative indices.

THEOREM 6.14 (**The argument principle**). *Let D be a domain in $\mathbb{C}$, and let $f \in \mathbf{M}(D)$. Suppose γ is a positively oriented Jordan curve in D that is homotopic to a point in D. Let $c \in \mathbb{C}$, and assume that $f(z) \neq c$ and $f(z) \neq \infty$ for $z \in \operatorname{range} \gamma$. Then*

$$\frac{1}{2\pi i} \int_\gamma \frac{f'(z)}{f(z) - c} \, dz = \sum_{z \in i(\gamma)} \nu_z(f - c) = Z - P,$$

where Z is the number of zeros of the function $f - c$ inside γ (counting multiplicities) and P is the number of poles of the function f inside γ (counting multiplicities).

PROOF. If $F(z) = \dfrac{f'(z)}{f(z) - c}$ for $z \in D$, then we claim that

$$\operatorname{Res}(F, \zeta) = \nu_\zeta(f - c) \text{ for all } \zeta \in D, \tag{6.1}$$

provided f is not identically c. The theorem then follows immediately from (6.1) and the Residue Theorem.

To verify our claim, it suffices to assume that $c = 0$ and that $\zeta = 0$. If $\nu_0(f) = n$, then $f(z) = z^n g(z)$ with g holomorphic near 0 and $g(0) \neq 0$. It follows that

$$f'(z) = nz^{n-1}g(z) + z^n g'(z)$$

and hence $\dfrac{f'(z)}{f(z)} = \dfrac{n}{z} + \dfrac{g'(z)}{g(z)}$. Thus $\dfrac{f'}{f}$ has residue n at 0. $\qquad \square$

REMARK 6.15. To explain the name of the theorem, we observe that the argument principle may be stated in the following way: Let D be a domain in $\mathbb{C}$, and let $f \in \mathbf{M}(D)$. Suppose γ is a positively oriented Jordan curve in D that is homotopic to a point in D. Let $c \in \mathbb{C}$, and suppose that $f(z) \neq c$ and $f(z) \neq \infty$ for all $z \in \operatorname{range} \gamma$.

If Z denotes the number of zeros of the function $f - c$ inside γ (counting multiplicities) and P denotes the number of poles of f inside γ (counting multiplicities), then the argument of $f - c$ increases by $2\pi(Z - P)$ upon traversing γ.

Indeed, note that $\dfrac{f'}{f - c} = (\log(f - c))'$ and recall that

$$\log(f - c) = \log|f - c| + i \arg(f - c).$$

Therefore,

$$\int_\gamma \frac{f'(z)}{f(z) - c} \, dz = \int_\gamma d\log|f(z) - c| + i \int_\gamma d\arg(f(z) - c)$$

The first integral on the rightmost side of the equation equals zero because $\log |f - c|$ is single-valued. The second integral on the rightmost side equals the change in the argument as one traverses γ.

COROLLARY 6.16. *Let f be a nonconstant holomorphic function on a neighborhood of $\zeta \in \mathbb{C}$, $\alpha = f(\zeta)$ and $m = \nu_\zeta(f - \alpha)$. Then there exist $r > 0$ and $\varepsilon > 0$ such that for all $\beta \in \mathbb{C}$ with $0 < |\beta - \alpha| < \varepsilon$, $f - \beta$ has exactly m simple zeros in $0 < |z - \zeta| < r$.*

PROOF. Observe that $m \geq 1$. Choose a positively oriented circle γ around ζ such that $f - \alpha$ vanishes only at ζ in $\operatorname{cl} i(\gamma)$ and $f'(z) \neq 0$ for all $z \in \operatorname{cl} i(\gamma) - \{\zeta\}$.

If we consider the curve $\gamma_1 = f \circ \gamma$, it follows from Theorem 6.14 that

$$I(\gamma_1, \alpha) = \frac{1}{2\pi i} \int_\gamma \frac{f'(z)}{f(z) - \alpha} \, dz = m\,.$$

Let $w = f(z)$, and conclude that for every β not in the range of γ_1 we have

$$\frac{1}{2\pi i} \int_\gamma \frac{f'(z)}{f(z) - \beta} \, dz = \frac{1}{2\pi i} \int_{\gamma_1} \frac{1}{w - \beta} \, dw = I(\gamma_1, \beta).$$

Now there exists a $\delta > 0$ such that $|f(z) - \alpha| \geq \delta$ for all $z \in \operatorname{range} \gamma$. Hence $|\beta - \alpha| < \delta$ implies that, for $z \in \gamma$,

$$|f(z) - \beta| = |(f(z) - \alpha) - (\alpha - \beta)| \geq |f(z) - \alpha| - |\alpha - \beta| > 0.$$

Thus $f - \beta$ does not vanish on γ for such β. Since $I(\gamma_1, \beta)$ is constant on each connected component of the complement of the range of γ_1 in $\mathbb{C}$, there is an $\varepsilon > 0$ such that

$$|\beta - \alpha| < \varepsilon < \delta \Rightarrow I(\gamma_1, \beta) = m.$$

If $\beta \neq \alpha$, then all the zeros of $f - \beta$ are simple (since f' is not zero close to ζ). Therefore $f - \beta$ has m simple zeros in $i(\gamma)$.   □

COROLLARY 6.17. *A nonconstant holomorphic function is an open mapping.*

PROOF. If $f : D \to \mathbb{C}$ is holomorphic on a domain D and is not a constant, we obtain from Corollary 6.16 that for any α in $f(D)$, there exists $\epsilon > 0$ such that $U(\alpha, \epsilon) \subseteq f(D)$, and the result follows.   □

COROLLARY 6.18. *An injective holomorphic function is a homeomorphism from its domain onto its image.*

Corollary 6.16 gives a characterization ($m = 1$ if and only if f is injective in a neighborhood of ζ) for a holomorphic function to be locally injective (or, equivalently, to be a local homeomorphism). See also the discussion in the next section.

THEOREM 6.19 (**Rouché's theorem**). *Let f and g be holomorphic functions on a domain D. Let γ be a positively oriented Jordan curve with $\operatorname{cl} i(\gamma)$ contained in D. Assume that $|f| > |g|$ on* range γ.
Then $Z_{f+g} = Z_f$, where Z_f denotes the number of zeros of f in $i(\gamma)$.

PROOF. Recall from Theorem 6.14 that $Z_f = I(f \circ \gamma, 0)$, and then use Theorem 4.51. □

THEOREM 6.20 (**Integral formula for the inverse function**). *Let $R > 0$. Suppose f is holomorphic on $|z| < R$, $f(0) = 0$, $f'(z) \neq 0$ for $|z| < R$, and $f(z) \neq 0$ for $0 < |z| < R$.*
For any $0 < r < R$, let γ_r be the positively oriented circle of radius r about 0, and let $m = \min |f|$ on γ_r.
Then

$$g(w) = \frac{1}{2\pi i} \int_{\gamma_r} \frac{t f'(t)}{f(t) - w} \, dt$$

defines a holomorphic function in $|w| < m$ with

$$f(g(w)) = w \quad on \ |w| < m$$

and

$$g(f(z)) = z \quad for \ z \in i(\gamma_r) \cap f^{-1}(|w| < m).$$

PROOF. Observe that $m > 0$, and fix w_0 with $|w_0| < m$. Then on the circle γ_r, we have

$$|f(z)| \geq m > |w_0|.$$

Thus f and $(f - w_0)$ have the same number of zeros in $i(\gamma_r)$, by Rouché's theorem, and hence $f(z) - w_0 = 0$ has a unique solution z_0 in $i(\gamma_r)$.
Therefore it suffices to show the following.

(1) $g(w_0) = z_0$ if $f(z_0) = w_0$, and

(2) g is a holomorphic function on the disk $|w| < m$.

To verify (1), note that it follows from the Residue Theorem that

$$g(w_0) = \operatorname{Res}(F, z_0), \quad where \ F(s) = \frac{s f'(s)}{f(s) - w_0} \quad for \ |s| < R.$$

Thus

$$g(w_0) = \lim_{s \to z_0} (s - z_0) \frac{s \, f'(s)}{f(s) - f(z_0)}$$
$$= \lim_{s \to z_0} \frac{(2s - z_0) f'(s) + (s^2 - s z_0) f''(s)}{f'(s)} = z_0,$$

where the second equality follows from l'Hopitals's rule (see Exercise 3.26).

Alternatively, to avoid the use of l'Hopitals's rule, we change the previous series of equalities to

$$g(w_0) = \lim_{s \to z_0} (s - z_0) \frac{s \, f'(s)}{f(s) - f(z_0)}$$
$$= \lim_{s \to z_0} (s - z_0) \frac{s f'(s)}{(s - z_0)(f'(z_0) + \frac{f''(z_0)}{2}(s - z_0) + \ldots)} = z_0.$$

To show (2) we note that $|f(t)| > |w|$ on γ_r and hence

$$\frac{1}{f(t) - w} = \frac{1}{f(t) \left[1 - \dfrac{w}{f(t)} \right]} = \frac{1}{f(t)} \sum_{n=0}^{\infty} \left(\frac{w}{f(t)} \right)^n.$$

Thus

$$g(w) = \frac{1}{2\pi i} \sum_{n=0}^{\infty} w^n \int_{\gamma_r} \frac{t \, f'(t)}{f(t)^{n+1}} \, dt.$$

Since $\left| \displaystyle\int_{\gamma_r} \frac{t \, f'(t)}{f(t)^{n+1}} \, dt \right| \le \dfrac{M}{m^{n+1}}$ for some constant M that is independent of n, the last power series has radius of convergence $\ge m$.   □

6.4. Local properties of holomorphic maps

Let D be a domain, $f \in \mathbf{H}(D)$, and $z_0 \in D$. In this section we describe the behavior of f near z_0 using results from the previous section. We will use the following notation:

$$z = x + iy, \ w = s + it = f(z) = u(x,y) + iv(x,y) \ \text{ for } z \in D,$$

and $w_0 = f(z_0)$.

PROPOSITION 6.21. *Let D be a domain in $\mathbb{C}$, z_0 a point in D and f a function holomorphic on D.*
   Then the following properties hold:
   (1) If $f'(z_0) \ne 0$, then f defines a homeomorphism of some neighborhood of z_0 onto some neighborhood of w_0.

PROOF. The condition implies that $\nu_{z_0}(f(z) - w_0) = 1$, and it follows from Corollary 6.16 that there exist $r > 0$ and $\varepsilon > 0$ such that for all $w \in \mathbb{C}$ with $0 < |w - w_0| < \varepsilon$, $f - w_0$ has exactly one simple zero in $0 < |z - z_0| < r$. In other words, f is injective near z_0. Now use Corollary 6.18 to conclude. $\square$

(2) *If there exists $n \in \mathbb{Z}_{\geq 1}$ such that*

$$0 = f'(z_0) = \ldots = f^{(n)}(z_0) \quad and \quad f^{(n+1)}(z_0) \neq 0,$$

then f is $n + 1$ to 1 near z_0.

PROOF. Let $g(z) = f(z) - w_0$. It is enough to prove that g is $n + 1$ to 1 near z_0. But

$$g(z_0) = 0 = g'(z_0) = \ldots = g^{(n)}(z_0), \quad and \quad g^{(n+1)}(z_0) \neq 0,$$

and therefore, for $|z - z_0|$ small, we may write

$$g(z) = \sum_{m \geq n+1} a_m (z - z_0)^m \quad (\text{where } a_{n+1} \neq 0)$$

$$= (z - z_0)^{n+1} \sum_{k=0}^{\infty} a_{k+n+1}(z - z_0)^k$$

$$= (z - z_0)^{n+1}(h(z))^{n+1} = (g_1(z))^{n+1},$$

where h and g_1 are holomorphic functions near z_0 such that $h(z_0) = (a_{n+1})^{\frac{1}{n+1}} \neq 0$, $g_1(z_0) = 0$, and $(g_1)'(z_0) = h(z_0) \neq 0$. The existence of h is a consequence of Exercise 5.1.

By (1), g_1 is a homeomorphism from a neighborhood of z_0 to a neighborhood of 0. Since $p(z) = z^{n+1}$ is clearly $n + 1$ to 1 from a neighborhood of 0 to a neighborhood of 0, and since $g = p \circ g_1$, it follows that g is $n + 1$ to 1 from a neighborhood of z_0 to a neighborhood of $g(z_0) = 0$, as claimed. $\square$

REMARK 6.22. The above property (2) of holomorphic mappings is also a consequence of Corollary 6.16. Much of the above discussion, as well as the next corollary, are slight amplifications of the material in the previous section.

An immediate consequence of (1) and (2) is the following

COROLLARY 6.23. *A holomorphic function f is injective near a point z_0 in its domain if and only if $f'(z_0) \neq 0$ if and only if f is a homeomorphism near z_0.*

(3) If $f'(z_0) \neq 0$, then angles between tangent vectors to curves at z_0 are preserved, and infinitesimal lengths at z_0 are multiplied by $|f'(z_0)|$.

If $0 = f'(z_0) = \ldots = f^n(z_0)$ and $f^{(n+1)}(z_0) \neq 0$ for some n in $\mathbb{Z}_{\geq 1}$, then angles between tangent vectors to curves at z_0 are multiplied by $n + 1$.

PROOF. Let us write

$$f'(z_0) = \lim_{z \to z_0} \frac{f(z) - f(z_0)}{z - z_0} = \lim_{z \to z_0} \frac{\Delta w}{\Delta z}.$$

Assume first that $f'(z_0) \neq 0$. Then

$$f'(z_0) = \rho\, e^{i\theta}, \quad \rho > 0.$$

If $z : [0, 1] \to D$ is a C^1-curve with $z(0) = z_0$ and $z'(0) \neq 0$, then $w = f \circ z$ is a C^1-curve with $w(0) = w_0$, $w'(0) \neq 0$.

Furthermore, if we denote $\Delta z = z - z_0$ (for z close to but different from z_0) and $\Delta w = f(z) - w_0$, then

$$\Delta w = \Delta z \frac{\Delta w}{\Delta z}$$

implies that

$$\arg \Delta w = \arg \Delta z + \arg \frac{\Delta w}{\Delta z},$$

which together with

$$\lim_{z \to z_0} \arg \frac{\Delta w}{\Delta z} = \theta$$

imply that

$$\arg w'(0) = \arg z'(0) + \arg f'(z_0).$$

All uses of the multivalued arg function need to be interpreted appropriately; we leave it to the reader to do so.

The assertion about lengths means that the ratio of the length of Δw to the length of Δz tends to $|f'(z_0)|$ as z tends to z_0. This follows immediately from

$$\lim_{z \to z_0} \left| \frac{\Delta w}{\Delta z} \right| = \rho.$$

The argument for the case with vanishing derivatives is almost identical to the one used to establish (2) and is hence left to the reader. $\qquad\square$

(4) Conversely, if $g \in \mathbf{C}^1(D)$ preserves angles, then $g \in \mathbf{H}(D)$.

PROOF. Let $z : [0, 1] \to D$ be a C^1-curve with $z'(t) \neq 0$ for all t. Then

$$w = g \circ z : [0, 1] \to g(D)$$

is also a C^1-curve and

$$w'(t) = g_z \, z'(t) + g_{\bar{z}} \, \overline{z'(t)} \ .$$

Since g preserves angles, $\arg \dfrac{w'(t)}{z'(t)}$ must be independent of $\arg z'(t)$. But

$$\frac{w'(t)}{z'(t)} = g_z + g_{\bar{z}} \frac{\overline{z'(t)}}{z'(t)}$$

and therefore $g_{\bar{z}} \equiv 0$.   □

(5) *The change in infinitesimal areas is given by multiplication by* $|f'(z_0)|^2$.

PROOF. We compute the Jacobian of the map f:

$$J(f) = \begin{vmatrix} u_x & v_y \\ v_x & v_y \end{vmatrix} = u_x v_y - u_y v_x = u_x^2 + v_x^2 = |f'(z_0)|^2 .$$

□

6.5. Evaluation of definite integrals

The Residue Theorem is a powerful tool for the evaluation of many definite integrals. We illustrate this with a few examples.

(1) The first integral to be evaluated is

$$\int_{-\infty}^{\infty} \frac{1}{x^4 + 1} \, dx \ .$$

We will obviously want to integrate $F(z) \, dz = \dfrac{1}{z^4 + 1} \, dz$. To apply the Residue Theorem, we must choose the path of integration carefully.

Let $R > 1$ and we choose γ_R to be the portion on $\mathbb{R}$ from $-R$ to $+R$ followed by the upper half of the circle $|z| = R$, as in Figure 6.1.

Since

$$z^4 + 1 = (z - e^{\frac{\pi i}{4}})(z - e^{\frac{3\pi i}{4}})(z + e^{\frac{\pi i}{4}})(z + e^{\frac{3\pi i}{4}}),$$

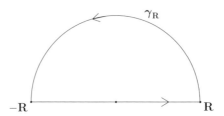

FIGURE 6.1. The path of integration for Example (1)

the function F has (possibly) nonzero residues only at these four roots of unity and we conclude that

$$\int_{\gamma_R} \frac{1}{z^4 + 1}\, dz = 2\pi\imath \left(\mathrm{Res}\left(F, e^{\frac{\pi\imath}{4}}\right) + \mathrm{Res}\left(F, e^{\frac{3\pi\imath}{4}}\right) \right).$$

The residues are easy to compute:

$$\mathrm{Res}(F, e^{\frac{\pi\imath}{4}}) = \frac{1}{(z^2 + \imath)(z + e^{\frac{\pi\imath}{4}})}\bigg|_{z=e^{\frac{\pi\imath}{4}}} = \frac{-1}{2(\sqrt{2} - \imath\sqrt{2})}$$

and

$$\mathrm{Res}(F, e^{\frac{3\pi\imath}{4}}) = \frac{1}{(z^2 - \imath)(z + e^{\frac{3\pi\imath}{4}})}\bigg|_{z=e^{\frac{3\pi\imath}{4}}} = \frac{1}{2(\sqrt{2} + \imath\sqrt{2})}.$$

Next we estimate the absolute value of the integral over the semicircle $\{z;\ \Im z \geq 0,\ |z| = R\}$:

$$\left| \int_0^\pi \frac{R\imath e^{\imath\theta}}{R^4 e^{4\imath\theta} + 1}\, d\theta \right| \leq \frac{\pi R}{R^4 - 1} \to 0 \ \text{as } R \to +\infty.$$

We conclude that $\int_{-\infty}^{\infty} \frac{1}{x^4 + 1}\, dx = \frac{\sqrt{2}\,\pi}{2}$.

REMARK 6.24. This method will work for the evaluation of integrals of the form $\int_{-\infty}^{+\infty} Q(x)\, dx$, where Q is a rational function with no singularities on $\mathbb{R}$ and with $\nu_\infty(Q) \geq 2$.

(2) A second class of integrals that can be evaluated by the Residue Theorem consists of those of the form

$$I = \int_0^{2\pi} Q(\cos\theta, \sin\theta)\, d\theta,$$

where Q is a rational function of two variables with no singularities on the unit circle $S^1 = \{z; |z| = 1\}$.

To apply the Residue Theorem, we express I as an integral of a holomorphic function over the unit circle. We use the change of variables

$$z = e^{i\theta}, \text{ hence } dz = e^{i\theta}\imath \, d\theta = \imath z \, d\theta$$

and

$$\cos\theta = \frac{e^{i\theta} + e^{-i\theta}}{2} = \frac{z + z^{-1}}{2}, \quad \sin\theta = \frac{e^{i\theta} - e^{-i\theta}}{2\imath} = \frac{z - z^{-1}}{2\imath}.$$

EXAMPLE 6.25. Let $0 < b < a$, and evaluate

$$I = \int_0^{2\pi} \frac{1}{a + b\cos\theta} \, d\theta = \int_{|z|=1} \frac{1}{(\imath z)\left(a + b\frac{z + \frac{1}{z}}{2}\right)} \, dz$$

$$= \int_{|z|=1} \frac{-2\imath}{bz^2 + 2az + b} \, dz$$

$$= 2\pi\imath \sum_{|z|<1} \operatorname{Res}\left(\frac{-2\imath}{bz^2 + 2az + b}, z\right).$$

The denominator of the integrand in the last integral is a quadratic polynomial in z with precisely one root inside the unit circle (the product of the roots is $+1$). We conclude that

$$I = 2\pi(a^2 - b^2)^{-\frac{1}{2}}.$$

(3) The last of the types of integrals to be discussed here is

$$I = \int_{-\infty}^{\infty} Q(x)e^{\imath x} dx,$$

where Q is a rational function with (at least) a simple zero at infinity and, in general, with no singularities on $\mathbb{R}$.

We illustrate with a more complicated example, where Q has a simple pole at the origin. Here the ordinary integral is replaced by its principal value (pr. v.) defined below.

$$\text{pr. v.} \int_{-\infty}^{\infty} \frac{e^{\imath x}}{x} dx = \lim_{\substack{\delta \to 0^+ \\ R_1 \to +\infty \\ R_2 \to +\infty}} \left(\int_{\delta}^{R_2} + \int_{-R_1}^{-\delta}\right) \frac{e^{\imath x}}{x} dx.$$

We must choose a nice contour for integration; start with large X_1, X_2, and Y and small δ, all positive. Our closed path γ has several segments (see Figure 6.2):

γ_1 : from $-X_1$ to $-\delta$ on $\mathbb{R}$,

γ_2 : the semicircle in the lower half-plane of radius δ and center 0,

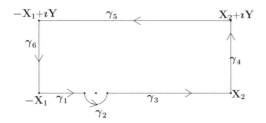

FIGURE 6.2. The path of integration for Example (3)

$$\gamma_3 : \text{from } \delta \text{ to } X_2 \text{ on } \mathbb{R},$$
$$\gamma_4 : \text{at } x = X_2 \text{ go up to height } Y,$$
$$\gamma_5 : \text{at height } Y \text{ travel from } X_2 \text{ back to } -X_1, \text{ and (finally)}$$
$$\gamma_6 : \text{at } x = -X_1 \text{ go down from height } Y \text{ to the real axis.}$$

We start with

$$\int_\gamma \frac{e^{\imath z}}{z}\, dz = 2\pi\imath \operatorname{Res}(f, 0)\,,$$

where $f(z) = \frac{e^{\imath z}}{z} = \frac{1}{z} + g(z)$, with g entire. Thus

$$\int_\gamma \frac{e^{\imath z}}{z}\, dz = 2\pi\,\imath\,.$$

We now estimate the integral over γ_4:

$$\left| \int_0^Y \frac{e^{\imath(X_2+\imath y)}}{X_2 + \imath y}\, \imath\, dy \right| \le \int_0^Y e^{-y} \frac{1}{|X_2 + \imath y|}\, dy$$
$$\le \frac{1}{X_2} \int_0^Y e^{-y}\, dy < \frac{1}{X_2}.$$

Next we estimate the integral over γ_5:

$$\left| \int_{X_2}^{-X_1} \frac{e^{\imath(x+\imath Y)}}{x + \imath Y}\, dx \right| \le \int_{-X_1}^{X_2} \frac{e^{-Y}}{|x + \imath Y|}\, dx$$
$$\le e^{-Y} \int_{-X_1}^{X_2} \frac{1}{Y}\, dx = \frac{e^{-Y}}{Y}[X_2 + X_1].$$

Also the integral over γ_6:

$$\left| \int_Y^0 \frac{e^{\imath(X_1+\imath y)}}{X_1 + \imath y}\, \imath\, dy \right| < \frac{1}{X_1}\,.$$

We conclude that

$$\int_\gamma \frac{e^{iz}}{z}\,dz = \lim_{\substack{\delta\to 0^+ \\ X_1\to+\infty \\ X_2\to\infty}} \int_{\gamma_1\cup\gamma_2\cup\gamma_3} \frac{e^{iz}}{z}\,dz.$$

Finally,

$$\lim_{\delta\to 0^+}\left(\int_\delta^{X_2} + \int_{-X_1}^{-\delta}\right)\frac{e^{ix}}{x}\,dx$$

$$= \lim_{\delta\to 0^+}\left(\int_{\gamma_1\cup\gamma_2\cup\gamma_3} \frac{e^{iz}}{z}\,dz + \int_{\gamma_2^-} \frac{e^{iz}}{z}\,dz\right).$$

But

$$\lim_{\delta\to 0^+}\int_{\gamma_2^-}(z^{-1} + g(z))\,dz = \lim_{\delta\to 0^+}\int_{\gamma_2^-} z^{-1}dz$$

because g is bounded on a neighborhood of 0 and the length of the path of integration goes to zero. Now

$$\lim_{\delta\to 0^+}\int_{\gamma_2^-} z^{-1}\,dz = \lim_{\delta\to 0^+}\int_0^{-\pi} \frac{1}{\delta e^{i\theta}}\delta e^{i\theta}\,i\,d\theta = -\pi i.$$

We conclude that

$$\text{pr. v. }\int_{-\infty}^\infty \frac{e^{ix}}{x}\,dx = \pi i.$$

Using the fact that $e^{ix} = \cos x + i\sin x$, we see that we have evaluated two real integrals:

$$\text{pr. v. }\int_{-\infty}^\infty \frac{\cos x}{x}\,dx = 0 \text{ and } \int_0^\infty \frac{\sin x}{x}\,dx = \frac{\pi}{2}.$$

Exercises

6.1. Use Rouché's theorem to prove the Fundamental Theorem of Algebra.

6.2. Let g be a holomorphic function on $|z| < R$, $R > 1$, with $|g(z)| \le 1$ for all $|z| \le R$.
  (1) Show that for all $t \in \mathbb{C}$ with $|t| < 1$, the equation

$$z = tg(z)$$

  has a unique solution $z = s(t)$ in the disk $|z| < 1$.
  (2) Show that $t \mapsto s(t)$ is a holomorphic function on the disk $|t| < 1$.

6.3. Verify (6.1) using Laurent series expansions for f and F.

6.4. Evaluate $\int_{-\infty}^{\infty} \dfrac{x}{4 + x^4}\, dx.$

6.5. If f is a holomorphic function on $0 < |z| < 1$ and f does not assume any value w with $|w - 1| < 2$, what can you conclude?

6.6. Compute $\int_{-\infty}^{\infty} \dfrac{dx}{1 + x^6}.$

6.7. Evaluate the following integrals.

(a) $\int_{-\infty}^{\infty} \dfrac{(x + 1)}{x^4 + 1}\, dx,$

(b) $\int_{0}^{\pi} \dfrac{d\theta}{\sqrt{5} + \cos \theta},$

(c) $\int_{|z|=1} \dfrac{z^6 dz}{7z^7 - 1},$

(d) $\int_{|z-100\pi|=\frac{199}{2}\pi} z \cot z\, dz.$

6.8. Let f be an entire function such that $|f(z)| = 1$ for $|z| = 1$. Which are the possible values for $f(0)$ and for $f(17)$?

6.9. Find $\int_{0}^{\infty} \dfrac{dx}{1 + x^3}$ using residues.

6.10. Find all functions f which are meromorphic in a neighborhood of $\{|z| \leq 1\}$ and such that $|f(z)| = 1$ for $|z| = 1$, f has a double pole at $z = \frac{1}{2}$, a triple zero at $z = -\frac{1}{3}$, and no other zeros or poles in $\{|z| < 1\}$.

6.11. Suppose f is an entire function satisfying $f(n) = n^5$ and $f\left(-\frac{n}{2}\right) = n^7$ for all $n \in \mathbb{Z}_{>0}$. How many zeros does the function $g(z) = [f(z) - e][f(z) - \pi]$ have?

6.12. Evaluate

$$\int_{|z|=3} \dfrac{f'(z)}{f(z) - 1}\, dz,$$

where $f(z) = 2 - 2z + z^2 + \dfrac{z^3}{81}.$

6.13. Suppose f is holomorphic for $|z| < 1$ and $f\left(\frac{1}{n}\right) = \frac{7}{n^3}$ for $n = 2, 3, \ldots$. What can be said about $f'''(0)$?

6.14. Let f be an entire function such that $|f(z)| \leq |z|^{\frac{23}{3}}$ for all $|z| > 10$. Compute $f^{(8)}(10.001)$.

6.15. Evaluate $\int_{|z-\frac{\pi}{2}|=3.15} z \tan z\, dz.$

6.16. Evaluate the following real integrals using residues:

$$\int_{-\infty}^{\infty} \frac{\cos x}{1+x^2} dx \ , \qquad \int_{-\infty}^{\infty} \frac{\sin x}{1+x^2} dx \ .$$

6.17. Find **all** Laurent series of the form $\sum_{-\infty}^{\infty} a_n z^n$ for the function

$$f(z) = \frac{z^2}{(1-z)^2(1+z)}.$$

6.18. If f is an entire function such that $\Re f(z) > -2$ for all $z \in \mathbb{C}$ and $f(\imath) = \imath + 2$, what is $f(-\imath)$?

6.19. If f is holomorphic on $0 < |z| < 2$ and satisfies $f(\frac{1}{n}) = n^2$ and $f(-\frac{1}{n}) = n^3$ for all $n \in \mathbb{Z}_{>0}$, what kind of singularity does f have at 0?

6.20. Let D be an open, bounded, and connected subset of $\mathbb{C}$ with smooth boundary.

If f is a nonconstant holomorphic function in a neighborhood of the closure of D such that $|f| = c$ is constant on ∂D, show that f takes on each value a such that $|a| < c$ at least once in D.

6.21. Suppose f is holomorphic in a neighborhood of the closure of the unit disk.

Show that for $|z| \leq 1$

$$f(z)(1 - |z|^2) = \frac{1}{2\pi\imath} \int_{|\tau|=1} \frac{1 - \bar{z}\tau}{\tau - z} f(\tau)\, d\tau$$

and conclude that the following inequality holds:

$$|f(z)|(1 - |z|^2) \leq \frac{1}{2\pi} \int_0^{2\pi} \left| f(\exp^{\imath\theta}) \right| d\theta \ .$$

6.22. Let f be an entire function. Suppose that $|f(z)| \leq A + B|z|^{10}$ for all $z \in \mathbb{C}$. Show that f is a polynomial.

6.23. Suppose f is meromorphic in a neighborhood of the closed unit disk, that it does not have zeroes nor poles in the unit disk, and that $|f(z)| = 1$ for $|z| = 1$. Find the most general such function.

6.24. Let C denote the positively oriented unit circle. Consider the function

$$f(z) = \frac{2\,z^{26}}{81} + \exp\left(z^{21}\right)\left(z - \frac{1}{2}\right)^2\left(z - \frac{1}{3}\right)^3.$$

Evaluate the following integrals:

$$\int_C f(z)\, dz\,; \qquad \int_C f'(z)\, dz\,; \qquad \int_C \frac{f'(z)}{f(z)}\, dz.$$

6.25. If f is entire and satisfies $|f''(z) - 3| \geq 0.001$ for all $z \in \mathbb{C}$, $f(0) = 0$, $f(1) = 2$, $f(-1) = 4$, what is $f(\imath)$?.

6.26. If f is holomorphic for $0 < |z| < 1$ and $f\left(\frac{1}{n}\right) = n^2$, $f\left(-\frac{1}{n}\right) = 2n^2$ for $n = 2, 3, 4, \ldots$, what can you say about f?

6.27. Find all series of the form $\sum_{-\infty}^{\infty} a_n z^n$ that converge in some domain to

$$f(z) = \frac{2 - z^2}{z(1 - z)(2 - z)}.$$

6.28. Suppose f is entire and $f(z) \neq t^2$ for all $z \in \mathbb{C}$ and for all $t \in \mathbb{R}$. Show that f is constant.

6.29. Find all Laurent series of the form $\sum_{-\infty}^{\infty} a_n z^n$ representing the function

$$f(z) = \frac{1}{(z - 1)(z - 2)(z - 3)}.$$

6.30. If f is holomorphic for $0 < |z| < 1$, $f\left(\frac{1}{n}\right) = n^{-2}$ and $f\left(-\frac{1}{2}\right) = 2n^{-2}$, $n = 2, 3, 4, \ldots$, find $\lim_{z \to 0} |f(z) - 2|$.

6.31. Find $\int_0^\infty \frac{\sin^2 2x}{x^2}\, dx$ using residues.

6.32. Prove the following extension of the Maximum Modulus Principle. Let f be holomorphic and bounded on $|z| < 1$, and continuous on $|z| \leq 1$ except maybe at $z = 1$. If $|f(e^{\imath\theta})| \leq A$ for $0 < \theta < 2\pi$, then $|f(z)| \leq A$ for all $|z| < 1$.

6.33. Let $\mathbb{D}$ denote the unit disk $\{z \in \mathbb{C}; |z| < 1\}$, and let $\{f_n\}$ be a sequence of holomorphic functions in $\mathbb{D}$ such that $\lim_{n \to \infty} f_n = f$ uniformly on compact subsets of $\mathbb{D}$.

Suppose that each f_n takes on the value 0 at most seven times on $\mathbb{D}$ (counted with multiplicity). Prove that either $f \equiv 0$ or f takes on the value 0 at most seven times on $\mathbb{D}$ (counted with multiplicity).

6.34. Show that the function $f(z) = z + 2z^2 + 3z^3 + 4z^4 + \cdots$ is injective in the unit disk $\mathbb{D} = \{z \in \mathbb{C}; |z| < 1\}$. Find $f(\mathbb{D})$.

6.35. Suppose f is a nonconstant function holomorphic on $\{z \in \mathbb{C}; |z| < 1\}$ and continuous on $\{z \in \mathbb{C}; |z| \leq 1\}$ such that for all $\theta \in \mathbb{R}$, the value $f(e^{\imath\theta})$ is on the boundary of the triangle with vertices 0, 1, and $\imath$.

Is there a z_0 with $|z_0| < 1$ such that $f(z_0) = \frac{1}{10}(1 + \imath)$? Is there a z_0 with $|z_0| < 1$ such that $f(z_0) = \frac{1}{2}(1 + \imath)$?

6.36. Is there a function f holomorphic for $|z| < 1$ and continuous for $|z| \le 1$ that satisfies

$$f(e^{\imath\theta}) = \cos\theta + 2\imath \sin\theta, \quad \text{for all } \theta \in \mathbb{R}?$$

CHAPTER 7

Sequences and Series of Holomorphic Functions

We now turn from the study of a single holomorphic function to the investigation of collections of holomorphic functions. In the first section we will see that under the appropriate notion of convergence of a sequence of holomorphic functions, the limit function inherits several properties that the approximating functions have, such as being holomorphic, and in the second section, we show that the space of holomorphic functions on a domain can be given the structure of a complete metric space. We then apply these ideas to obtain a series expansion for the cotangent function. In the fourth section, we characterize the compact subsets of the space of holomorphic functions on a domain. This powerful characterization is used in Section 7.5 to study results on approximations of holomorphic functions and, in particular, to prove Runge's theorem. This characterization will also be used in Chapter 8 to prove the Riemann Mapping Theorem.

7.1. Consequences of uniform convergence on compact sets

We begin by recalling some notation and introducing some new symbols. Let D be a domain in $\mathbb{C}$. We denote by $\mathbf{C}(D)$ the vector space of continuous complex-valued functions on D and recall that $\mathbf{H}(D)$ is the vector space of holomorphic functions on D (see Definition 3.49). Note that $\mathbf{H}(D) \subset \mathbf{C}(D)$.

PROPOSITION 7.1. *A necessary and sufficient condition for a sequence of functions $\{f_n\} \subset \mathbf{C}(D)$ to converge uniformly on all compact subsets of D is for the sequence to converge uniformly on all compact disks with rational centers and rational radii contained in D.*

PROOF. Every compact set contained in D can be covered by finitely many such disks. □

It is clear that if a sequence of functions $\{f_n\} \subset \mathbf{C}(D)$ converges uniformly to a function f on all compact subsets of D, then for all z in D, we have $\lim_{n \to \infty} f_n(z) = f(z)$. The converse is not true: Uniform convergence on all compact subsets of D is stronger than pointwise convergence. To see this observe that we know from Theorem 2.23 that if a sequence of functions $\{f_n\} \subset \mathbf{C}(D)$ converges uniformly to a function f on all compact subsets of D, then $f \in \mathbf{C}(D)$. Consider any sequence of continuous functions converging at every point of the domain to a discontinuous function. Such an example is easily constructed (see Exercise 7.1).

We now study some consequences of this notion of uniform convergence on compact subsets of D, also called *locally uniform convergence*, for $\mathbf{H}(D)$. The first one says that $\mathbf{H}(D)$ is closed under this type of convergence.

THEOREM 7.2. *If $\{f_n\} \subset \mathbf{H}(D)$ and $\{f_n\}$ converges uniformly on all compact subsets of D, then the limit function f is holomorphic on D.*

PROOF. We already know that $f \in \mathbf{C}(D)$.

Let γ be any closed curve homotopic to a point in D. Then, by Cauchy's theorem,

$$\int_\gamma f_n(z)\, dz = 0.$$

By uniform convergence it follows that

$$\int_\gamma f(z)\, dz = \lim_{N \to \infty} \int_\gamma f_n(z)\, dz = 0,$$

and then, by Morera's theorem, f is holomorphic on D. □

COROLLARY 7.3. *If $\{f_n\} \subset \mathbf{H}(D)$ and $\sum_{n=1}^{\infty} f_n$ converges uniformly on all compact subsets of D, then the limit function (also denoted by $\sum_{n=1}^{\infty} f_n$) is holomorphic on D.*

Theorem 7.2 has no analog in real variables: It is easy to see (at least pictorially) that the absolute value function on $\mathbb{R}$, which has no derivative at 0, can be approximated uniformly by differentiable functions. A more extreme example was constructed by Weierstrass, that of a continuous function defined on $[0, 1]$ which is nowhere differentiable and uniformly approximated by polynomials.

The next consequence of uniform convergence on compact sets is that uniform convergence of a sequence of holomorphic functions on compact subsets implies uniform convergence of the derivatives on compact subsets. This is another feature of holomorphic functions not shared by real differentiable functions: It is easy to construct a sequence of differentiable functions converging uniformly on a closed interval and such that the sequence of derivatives does not converge uniformly. We leave this construction to the reader as Exercise 7.2.

THEOREM 7.4. *If $\{f_n\} \subset \mathbf{H}(D)$ and $f_n \to f$ uniformly on all compact subsets of D, then $f_n' \to f'$ uniformly on all compact subsets of D.*

PROOF. Since $f \in \mathbf{H}(D)$, it is enough to check uniform convergence of the derivatives on small compact subdisks $R \subset D$ with $\partial R = \gamma$ positively oriented. For $z \in i(\gamma)$, we have

$$f'(z) = \frac{1}{2\pi i} \int_\gamma \frac{f(\zeta)}{(\zeta - z)^2} \, d\zeta = \lim_{n \to \infty} \frac{1}{2\pi i} \int_\gamma \frac{f_n(\zeta)}{(\zeta - z)^2} \, d\zeta = \lim_{n \to \infty} f_n'(z);$$

this convergence is uniform in any smaller compact subdisk, such as

$$\widetilde{R} = \{z \in i(\gamma); \inf\{|z - w| \, ; w \in \partial R\} \geq \delta > 0\},$$

with δ sufficiently small. □

THEOREM 7.5. *Let $\{f_n\}$ be a sequence of holomorphic functions on D such that $f_n \to f$ uniformly on all compact subsets of D.*
If $f_n(z) \neq 0$ for all $z \in D$ and all $n \in \mathbb{Z}_{>0}$, then either

(a) f is identically zero, or
(b) $f(z) \neq 0$ for all $z \in D$.

PROOF. Assume that there is a $\zeta \in D$ with $f(\zeta) = 0$ and that f is not identically zero. Then there exists a circle γ with center ζ such that $\mathrm{cl}\,i(\gamma) \subset D$ and $f(z) \neq 0$ for all $z \in \mathrm{cl}\,i(\gamma) - \{\zeta\}$.
Therefore the number N of zeros of f in $i(\gamma)$ satisfies

$$N = \frac{1}{2\pi i} \int_\gamma \frac{f'(z)}{f(z)} \, dz \geq 1.$$

But

$$\int_\gamma \frac{f'(z)}{f(z)} \, dz = \lim_{n \to \infty} \int_\gamma \frac{f_n'(z)}{f_n(z)} \, dz = 0.$$

□

DEFINITION 7.6. Let $f \in \mathbf{H}(D)$. We call f *simple, univalent,* or *schlicht* if it is one-to-one (injective) on D.

REMARK 7.7. A schlicht function is a homeomorphism with nowhere-vanishing derivative.

THEOREM 7.8 (**Hurwitz**). *Assume D is a domain in $\mathbb{C}$. If $\{f_n\}$ is a sequence in $\mathbf{H}(D)$ with $f_n \to f$ uniformly on all compact subsets of D and f_n is schlicht for each n, then either f is constant or schlicht.*

PROOF. Assume that f is neither constant nor schlicht; thus, in particular, there exist z_1 and z_2 in D with $z_1 \neq z_2$ and $f(z_1) = f(z_2)$.

Look at $g_n(z) = f_n(z) - f_n(z_2)$ on the domain $D' = D - \{z_2\}$. Then $g_n \in \mathbf{H}(D')$, g_n never vanishes and $g_n \to g = f - f(z_2)$ uniformly on all compact subsets of D'. But g is not identically zero, and it vanishes at z_1.   □

7.2. A metric on $\mathbf{C}(D)$

We now introduce, for use in the proof of the Compactness Theorem of this chapter and in the proof of the Riemann Mapping Theorem in Chapter 8, a *metric* on $\mathbf{C}(D)$ for any domain D in $\mathbb{C}$. The metric ρ on $\mathbf{C}(D)$ will have the property that convergence in the ρ-metric is equivalent to uniform convergence on all compact subsets of D.

For K compact in D and $f \in \mathbf{C}(D)$, set

$$||f||_K = \max\{|f(z)| \, ; \; z \in K\}.$$

Consider the set of compact (closed) disks contained in D with rational radii and rational centers (that is, if the center of the disk is at $z = x + \imath y$, then both x and $y \in \mathbb{Q}$). There are countably many such disks and they cover D. Call this collection of disks $\{D_i\}_{i\in\mathbb{Z}_{>0}}$.

For $n \in \mathbb{Z}_{>0}$, let

(1)
$$K_n = \bigcup_{i \leq n} D_i,$$

then $\{K_n\}$ is an *exhaustion* of D; that is,

(2)
$$\text{each } K_n \text{ is compact,}$$

(3)
$$K_n \subset K_{n+1} \text{ for all } n \in \mathbb{Z}_{>0}, \quad \text{and}$$

(4)
$$\bigcup_{n\in\mathbb{Z}_{>0}} \text{int } K_n = D,$$

where int K denotes the interior of the set K.

From now on, we shall use only properties (2), (3) and (4) of our exhaustion and *not* how these sets were constructed.

REMARK 7.9. A crucial consequence of these properties that we will use often is that *each compact subset K of D is contained in K_n for some n.*

For $f \in \mathbf{C}(D)$ and $i \in \mathbb{Z}_{>0}$, we set

$$M_i(f) = ||f||_{K_i},$$

and we note that

$$M_{i+1} \geq M_i.$$

We define

$$d(f) = \sum_{i=1}^{\infty} 2^{-i} \min(1, M_i(f)) \leq \sum_{i=1}^{\infty} 2^{-i} = 1. \tag{7.1}$$

Properties of d. For all f and $g \in \mathbf{C}(D)$,

    (1) $d(f) \geq 0$, and $d(f) = 0$ if and only if $f \equiv 0$.

    (2) $d(f + g) \leq d(f) + d(g)$.

    (3) For each i, $2^{-i} \min(1, M_i(f)) \leq d(f)$.

    (4) For each i, $d(f) \leq M_i(f) + 2^{-i}$.

PROOF OF PROPERTY (2).

$$d(f+g) = \sum_{i=1}^{\infty} 2^{-i} \min(1, M_i(f+g)) \leq \sum_{i=1}^{\infty} 2^{-i} \min(1, M_i(f) + M_i(g))$$

$$\leq \sum_{i=1}^{\infty} 2^{-i} [\min(1, M_i(f)) + \min(1, M_i(g))].$$

                                                                     □

PROOF OF PROPERTY (4).

$$d(f) = \sum_{j \leq i} 2^{-j} \min(1, M_j(f)) + \sum_{j > i} 2^{-j} \min(1, M_j(f))$$

$$\leq \sum_{j \leq i} 2^{-j} M_j(f) + \sum_{j > i} 2^{-j} \leq \left(\sum_{j \leq i} 2^{-j} \right) M_i(f) + 2^{-i}.$$

                                                                     □

Finally, we define the *metric* on $\mathbf{C}(D)$ we have been seeking:

$$\rho(f, g) = d(f - g). \tag{7.2}$$

Properties of ρ. For all f, g, and h in $\mathbf{C}(D)$, the following hold:

(1) $\rho(f, g) \geq 0$, and $\rho(f, g) = 0$ if and only if $f = g$.

(2) $\rho(f, g) = \rho(g, f)$.

(3) $\rho(f, g) \leq \rho(f, h) + \rho(h, g)$.

(4) $\rho(f + h, g + h) = \rho(f, g)$; that is, ρ is translation invariant.

PROOF OF PROPERTY (3):

$$\rho(f, g) = d(f - g) = d(f - h + h - g) \leq d(f - h) + d(h - g).$$

□

Note that properties (1) to (3) say that ρ is a metric on $\mathbf{C}(D)$.

THEOREM 7.10. *Convergence in the ρ-metric in $\mathbf{C}(D)$ is equivalent to uniform convergence on all compact subsets of D.*

PROOF. Let $\{f_n\} \subset \mathbf{C}(D)$, and assume that $\{f_n\}$ is ρ-convergent. Since for every compact set $K \subset D$ there is an i such that $K \subset K_i$, it suffices to show uniform convergence on K_i for each i.

Given $0 < \epsilon < 1$, choose N large so that

$$\rho(f_m, f_n) = d(f_m - f_n) < 2^{-i}\epsilon \text{ for all } m, n \geq N.$$

Now

$$2^{-i}\min(1, M_i(f_m - f_n)) \leq d(f_m - f_n) < 2^{-i}\epsilon,$$

and thus

$$M_i(f_m - f_n) < \epsilon < 1;$$

that is,

$$\|f_m - f_n\|_{K_i} < \epsilon.$$

The above inequality implies that the sequence $\{f_n\}$ converges uniformly on K_i. If $\epsilon \geq 1$, then use $\epsilon_0 = \frac{3}{4}$ and proceed as above.

We have actually shown more than claimed: If $\{f_n\}$ is a ρ-Cauchy sequence in $\mathbf{C}(D)$, then there exists an $f \in \mathbf{C}(D)$ such that $f_n \to f$ uniformly on all compact subsets of D.

Conversely, assume that $f_n \to f$ uniformly on K_i for all i. Thus

$$\lim_{n \to \infty} M_i(f - f_n) = 0 \text{ for all } i.$$

Given $\epsilon > 0$, first choose i such that $2^{-i} < \frac{\epsilon}{2}$ and next choose N such that $M_i(f - f_n) < \frac{\epsilon}{2}$ for all $n \geq N$. Then

$$d(f - f_n) \leq M_i(f - f_n) + 2^{-i} < \frac{\epsilon}{2} + \frac{\epsilon}{2} = \epsilon.$$

□

COROLLARY 7.11. *The topology of the metric space* $(\mathbf{C}(D), \rho)$ *is independent of the choice of exhaustion* $\{K_n\}_{n \in \mathbb{Z}_{>0}}$ *of* D.

COROLLARY 7.12. ρ *is a complete metric on* $\mathbf{C}(D)$.

Because of Theorem 7.10, we can reformulate the results of the previous section in terms of the metric ρ. In particular, Theorems 7.2 and 7.4 can now be phrased as in the following Corollary. We already remarked that $\mathbf{H}(D) \subset \mathbf{C}(D)$. We let $\rho|_{\mathbf{H}(D)}$ denote the restriction of the metric ρ to $\mathbf{H}(D)$.

COROLLARY 7.13. $\mathbf{H}(D)$ *is a closed subspace of* $(\mathbf{C}(D), \rho)$. *As such,* $(\mathbf{H}(D), \rho|_{\mathbf{H}(D)})$ *is a complete metric space. Furthermore,* $f \mapsto f'$ *is a continuous function from* $\mathbf{H}(D)$ *to itself.*

It is useful to have an alternative description of the topology of $(\mathbf{C}(D), \rho)$. Toward this end, we make the following definition.

DEFINITION 7.14. Given $f \in \mathbf{C}(D)$, K compact $\subset D$ and $\epsilon > 0$, we define

$$N_f(\epsilon) = \{g \in \mathbf{C}(D);\ \rho(g, f) < \epsilon\}$$

and

$$V_f(K, \epsilon) = \{g \in \mathbf{C}(D);\ ||g - f||_K < \epsilon\}.$$

THEOREM 7.15. *For any* $f \in \mathbf{C}(D)$, *a basis for the neighborhood system at* f *is given by the sets* $V_f(K, \epsilon)$.

PROOF. We must show that

(1) given $V_f(K, \epsilon)$, there exists an $N_f(\delta) \subset V_f(K, \epsilon)$,

and

(2) given $N_f(\delta)$, there exists a $V_f(K, \epsilon) \subset N_f(\delta)$.

To show (1), we assume without loss of generality that $0 < \epsilon < 1$. Choose i such that $K \subset K_i$, and set $\delta = 2^{-i}\epsilon$. If $g \in N_f(\delta)$, then $d(g - f) < 2^{-i}\epsilon$. Thus

$$2^{-i}\min(1, M_i(g - f)) < 2^{-i}\epsilon$$

and then

$$M_i(g - f) = ||g - f||_{K_i} < \epsilon.$$

But

$$||g - f||_K \leq ||g - f||_{K_i};$$

that is, $g \in V_f(K, \epsilon)$.

To show (2), choose i such that $2^{-i} < \frac{\delta}{2}$. For $g \in V_f\left(K_i, \frac{\delta}{2}\right)$, we have

$$M_i(g - f) < \frac{\delta}{2}.$$

Hence

$$\rho(g, f) = d(g - f) < M_i(g - f) + 2^{-i} < \delta;$$

that is, $g \in N_f(\delta)$. □

REMARK 7.16. $f_n \to f$ in the ρ-metric if and only if for all compact $K \subset D$ and all $\epsilon > 0$, there exists $N = N(K, \epsilon)$ in $\mathbb{Z}_{>0}$ such that $\|f - f_n\|_K < \epsilon$ for all $n > N$.

We can apply these concepts to convergence of meromorphic functions.

DEFINITION 7.17. Let $\{f_n\}$ be a sequence in $\mathbf{M}(D)$, the meromorphic functions on D. We say that $\sum f_n$ converges uniformly (absolutely) on a subset A of D if there exists an integer N such that f_n is holomorphic on A for all $n > N$ and $\sum\limits_{N+1}^{\infty} f_n$ converges uniformly (absolutely) on A.

THEOREM 7.18. Let $\{f_n\} \subset \mathbf{M}(D)$. If $\sum f_n$ converges uniformly on compact subsets of D, then the series $f = \sum f_n$ is a meromorphic function on D and $\sum f'_n$ converges uniformly on all compact subsets to f'.

PROOF. The proof is trivial. □

7.3. The cotangent function

As an application of the ideas developed in the last two sections, we establish a series expansion formula for the cotangent function.

THEOREM 7.19. For all z in $\mathbb{C} - \mathbb{Z}$, the following equalities hold:

$$\pi \cot \pi z = \frac{\pi \cos \pi z}{\sin \pi z} = \frac{1}{z} + \sum_{n \in \mathbb{Z}, \, n \neq 0} \left[\frac{1}{z - n} + \frac{1}{n} \right] \qquad (7.3)$$

$$= \frac{1}{z} + 2z \sum_{n=1}^{\infty} \left[\frac{1}{z^2 - n^2} \right].$$

We first observe that the function $F(z) = \dfrac{\cos \pi z}{\sin \pi z}$ has its poles at the integers, each of these poles is simple and has residue 1. It would

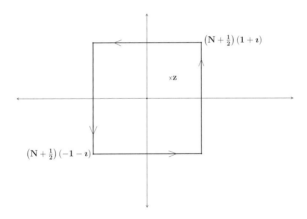

FIGURE 7.1. The square C_N

seem more natural to sum the series $\displaystyle\sum_{n=-\infty}^{\infty} \frac{1}{z-n}$, but this one does not converge (Exercise 7.3).

We claim that $\displaystyle\sum_{n\in\mathbb{Z},\, n\neq 0} \left[\frac{1}{z-n} + \frac{1}{n} \right] = \sum_{n\neq 0} \frac{z}{n(z-n)}$ converges absolutely and uniformly on compact subsets of $\mathbb{C}$.

To verify this claim, assume that $|z| \leq R$ with $R > 0$. Then

$$\sum_{|n|\geq 2R} \frac{|z|}{|n|\,|n-z|} \leq \sum_{|n|\geq 2R} \frac{R}{|n|\,(|n|-R)}$$

$$\leq \sum_{|n|\geq 2R} \frac{R}{|n|\,\left|\frac{n}{2}\right|} \leq 2R \sum_{n\neq 0} \frac{1}{|n|^2} < +\infty\ .$$

We can now verify the expansion (7.3) for $\pi \cot \pi z$.

PROOF OF THEOREM 7.19. For $N \in \mathbb{Z}_{>0}$, let C_N be the positively oriented boundary of the square with vertices $\left(N + \frac{1}{2}\right)(\pm 1 \pm \imath)$ (see Figure 7.1).

Then

$$\frac{1}{2\pi\imath} \int_{C_N} \frac{\cot \pi t}{t-z}\, dt = \sum_{t\in i(C_N)} \mathrm{Res}\left(\frac{\cot \pi t}{t-z}, t \right).$$

Here $z \in \mathbb{C}$ is fixed: We take $z \notin \mathbb{Z}$, $z \in i(C_N)$. The poles of the function $H(t) = \dfrac{\cot \pi t}{t-z}$ occur at $t = z$ and at $t = n \in \mathbb{Z}$, and they are all simple.

Furthermore, we see that

$$\operatorname{Res}\left(\frac{\cot \pi t}{t - z}, z\right) = \cot \pi z$$

and

$$\operatorname{Res}\left(\frac{\cot \pi t}{t - z}, n\right) = \lim_{t \to n}(t - n)\frac{\cos \pi t}{\sin \pi t}\frac{1}{t - z} = \frac{1}{\pi(n - z)}.$$

Thus we have

$$\frac{1}{2\pi i}\int_{C_N}\frac{\cot \pi t}{t - z}\,dt = \cot \pi z + \frac{1}{\pi}\sum_{n=-N}^{N}\frac{1}{n - z}$$

$$= \cot \pi z + \frac{1}{\pi}\sum_{\substack{n=-N \\ n\neq 0}}^{N}\left[\frac{1}{n - z} - \frac{1}{n}\right] - \frac{1}{\pi z},$$

where the last equality holds because

$$\sum_{\substack{n=-N \\ n\neq 0}}^{N}\frac{1}{n} = 0.$$

Hence it suffices to prove

LEMMA 7.20. *We have*

$$\lim_{N \to \infty}\int_{C_N}\frac{\cot \pi t}{t - z}\,dt = 0.$$

REMARK 7.21. We will also have obtained

$$\pi \cot \pi z = \frac{1}{z} - \sum_{n=1}^{\infty}\left[\frac{1}{n - z} - \frac{1}{n + z}\right] = \frac{1}{z} - \sum_{n=1}^{\infty}\frac{2z}{n^2 - z^2},$$

where the last series converges uniformly and absolutely on compact subsets of $\mathbb{C} - \mathbb{Z}$.

PROOF OF LEMMA. We proceed in stages:

(1) $\dfrac{1}{2\pi i}\displaystyle\int_{C_N}\frac{\cot \pi t}{t}\,dt = 0.$

As usual, for $G(t) = \dfrac{\cot \pi t}{t}$ we have

$$\frac{1}{2\pi i}\int_{C_N}\frac{\cot \pi t}{t}\,dt = \sum_{t\in i(C_N)}\operatorname{Res}(G, t) = \operatorname{Res}(G, 0) + \sum_{\substack{n=-N \\ n\neq 0}}^{N}\frac{1}{\pi n}.$$

The last sum is clearly zero, and the residue of G at 0 is 0 because G is an even function.

(2) $$\int_{C_N} \frac{\cot \pi t}{t - z} dt = \int_{C_N} \cot \pi t \left[\frac{1}{t - z} - \frac{1}{t} \right] dt = \int_{C_N} \cot \pi t \left[\frac{z}{t(t - z)} \right] dt,$$

where the first equality holds by (1).

(3) There exists an $M > 0$ (independent of N) such that
$$|\cot \pi t| \leq M \text{ on } C_N.$$

For $t = u + iv$,
$$|\cos \pi t|^2 = \cos^2 \pi u + \sinh^2 \pi v ,$$
$$|\sin \pi t|^2 = \sin^2 \pi u + \sinh^2 \pi v ,$$

and thus
$$|\cot \pi t|^2 = \frac{\cos^2 \pi u + \sinh^2 \pi v}{\sin^2 \pi u + \sinh^2 \pi v} .$$

On the vertical sides of C_N, we have $u = \pm N \pm \frac{1}{2}$, and hence
$$\cos^2 \pi u = \cos^2 \left(\pi \left(\pm N \pm \frac{1}{2} \right) \right) = 0,$$
$$\sin^2 \pi u = \sin^2 \left(\pi \left(\pm N \pm \frac{1}{2} \right) \right) = 1, \text{ and}$$
$$|\cot \pi t|^2 = \frac{\sinh^2 \pi v}{1 + \sinh^2 \pi v} \leq 1.$$

On the horizontal sides of C_N, $v = \pm N \pm \frac{1}{2}$, and hence
$$|\cot \pi t|^2 \leq \frac{1 + \sinh^2 \pi \left(\pm N \pm \frac{1}{2} \right)}{\sinh^2 \pi \left(\pm N \pm \frac{1}{2} \right)} \to 1 \text{ as } N \to \infty.$$

Thus, there exists an $M > 0$ such that $|\cot \pi t| \leq M$ for t on the horizontal sides and the claim is proved.

(4) If we denote by $L(C_N)$ the length of C_N, then
$$\left| \int_{C_N} \frac{\cot \pi t}{t - z} dt \right| = \left| \int_{C_N} \cot \pi t \left[\frac{z}{t(t - z)} \right] dt \right|$$
$$\leq \int_{C_N} \frac{M |z|}{|t| |t - z|} |dt| \leq \frac{M |z|}{\left| N + \frac{1}{2} \right| \left(\left| N + \frac{1}{2} \right| - |z| \right)} L(C_N)$$
$$= \frac{M |z|}{\left| N + \frac{1}{2} \right| \left(\left| N + \frac{1}{2} \right| - |z| \right)} 4 (2N + 1) \to 0 \text{ as } N \to \infty.$$

$\square$

Thus the theorem is proved.   □

Differentiating the series (7.3) term by term, we obtain the following expansion.

COROLLARY 7.22. *For all $z \in \mathbb{C} - \mathbb{Z}$,*

$$\frac{\pi^2}{\sin^2 \pi z} = \sum_{n=-\infty}^{\infty} \frac{1}{(z-n)^2}.$$

In particular,

$$\frac{\pi^2}{4} = \sum_{n=-\infty}^{\infty} \frac{1}{(2n-1)^2}.$$

7.4. Compact sets in $\mathbf{H}(D)$

We now return to the study of $\mathbf{C}(D)$ with the ρ-metric.

Recall that a metric space X is compact if and only if every sequence in X has a subsequence that converges to a point in X, and a subset X of $\mathbb{R}^n$ is compact if and only if it is closed and bounded. This last statement is the result we are trying to generalize to $\mathbf{H}(D)$.

DEFINITION 7.23. Let $A \subset \mathbf{C}(D)$. We say that A is *bounded in the strong sense* or *strongly bounded* if for all compact $K \subset D$ and all $\epsilon > 0$ there exists a $\lambda > 0$ such that

$$A \subset \lambda V_0(K, \epsilon) = \{g \in \mathbf{C}(D);\ g = \lambda f \text{ with } ||f||_K < \epsilon\}.$$

REMARK 7.24. In a metric space (X, ρ), one defines for any subset A of X

$$\operatorname{diam} A = \sup\{\rho(f, g);\ f \in A \text{ and } g \in A\}.$$

Usually one says that A is bounded if $\operatorname{diam} A < +\infty$. For a bounded metric (as in our case), this concept is not very useful.

LEMMA 7.25. *A set $A \subset \mathbf{C}(D)$ is strongly bounded if and only if for each compact $K \subset D$, there exists an $M(K) > 0$ such that $||f||_K \leq M(K)$ for all $f \in A$; that is, A is strongly bounded if and only if the functions in A are uniformly bounded on each compact subset of D.*

PROOF. We leave the proof as Exercise 7.4.   □

THEOREM 7.26. *A compact subset $A \subset \mathbf{C}(D)$ is closed and strongly bounded.*

PROOF. A compact subset of any metric space is closed. If $K \subset D$ is compact, then

$$f \mapsto \|f\|_K$$

is continuous on the compact set A and hence strongly bounded. $\qquad \square$

LEMMA 7.27. *Let $\zeta \in \mathbb{C}$ and $D = U(\zeta, R)$ for some $R > 0$. Let $A \subset \mathbf{H}(D)$ be strongly bounded, and let $\{f_k\}_{k=1}^{\infty} \subset A$.*

Then the sequence $\{f_k\}$ converges uniformly on all compact subsets of D if and only if $\lim_{k \to \infty} f_k^{(n)}(\zeta)$ exists (in $\mathbb{C}$) for all integers $n \geq 0$.

PROOF. If $f_k \to f$ uniformly on all compact subsets of D, then for every nonnegative integer n, $f_k^{(n)} \to f^{(n)}$ uniformly on compact subsets of D; in particular $f_k^{(n)}(\zeta) \to f^{(n)}(\zeta)$, as a set consisting of one point is certainly compact.

Conversely, it suffices to show that $\{f_k\}$ converges uniformly on $\mathrm{cl}\, U(\zeta, r)$ with $0 < r < R$. Choose r_0 such that $r < r_0 < R$. Since A is strongly bounded, there exists an $M = M(r_0)$ such that

$$|f_k(z)| \leq M \text{ for } |z - \zeta| \leq r_0.$$

Write

$$f_k(z) = \sum_{n \geq 0} a_{n,k}(z - \zeta)^n \text{ for } |z - \zeta| < R;$$

then Cauchy's inequalities (5.3) tell us that

$$|a_{n,k}| \leq \frac{M}{r_0^n} \text{ for all } n \text{ and } k.$$

Assume $|z - \zeta| \leq r$. Then

$$|f_k(z) - f_m(z)| \leq \left| \sum_{n=0}^{\infty} a_{n,k}(z - \zeta)^n - \sum_{n=0}^{\infty} a_{n,m}(z - \zeta)^n \right|$$

$$\leq \left| \sum_{n=0}^{N} (a_{n,k} - a_{n,m})(z - \zeta)^n \right| + 2M \sum_{n=N+1}^{\infty} \left(\frac{r}{r_0} \right)^n.$$

Let $\epsilon > 0$, and choose $N_0 \in \mathbb{Z}_{>0}$ such that

$$2M \sum_{n=N+1}^{\infty} \left(\frac{r}{r_0} \right)^n < \frac{\epsilon}{2} \text{ for } N > N_0.$$

Choose N_1 such that $k, m \geq N_1$ implies

$$\left| \sum_{n=0}^{N_0} (a_{n,k} - a_{n,m})(z - \zeta)^n \right| < \frac{\epsilon}{2}.$$

This may be achieved by requiring, for example, that

$$|a_{n,k} - a_{n,m}| < \frac{\epsilon}{2N_0 r_0^n} \ ;$$

this last finite set of inequalities can be satisfied because

$$\lim_{k,m\to\infty} |a_{n,k} - a_{n,m}| = 0 \ \text{ for each } n \ .$$

$\square$

THEOREM 7.28 (**Compactness Theorem**). *Let D be a domain in $\mathbb{C}$. Then every closed subset A of $\mathbf{H}(D)$ that is bounded in the strong sense is compact.*

PROOF. Cover D by countably many open disks $\{U(z_i, r_i)\}_{i\in\mathbb{Z}_{>0}}$ whose closures are contained in D.

For each $i \in \mathbb{Z}_{>0}$ and each $n \in \mathbb{Z}_{\geq 0}$, consider the mapping

$$\lambda_i^n : \mathbf{H}(D) \to \mathbb{C}, \ \lambda_i^n(f) = f^{(n)}(z_i).$$

The maps λ_i^n are $\mathbb{C}$-linear and continuous.

Given a sequence $\{f_k\}_{k=1}^{\infty} \subset A$, we consider the set of numbers

$$\lambda_i^n(f_k) = f_k^{(n)}(z_i).$$

We show that there exists $B \subset \mathbb{Z}_{>0}$, $|B| = \infty$, such that

$$\lim_{\substack{k\in B \\ k\to+\infty}} f_k^{(n)}(z_i) \text{ exists for all } n \text{ and } i. \tag{7.4}$$

Assertion (7.4) suffices to prove the theorem; for then by Lemma 7.27, the sequence $\{f_k\}_{k\in B}$ converges uniformly on the closed disk $\operatorname{cl} U(z_i, r_i)$ for each i, which implies that the same sequence converges uniformly on all compact subsets of D. Since A is closed, $\lim_{\substack{k\in B \\ k\to+\infty}} f_k \in A$. Thus every sequence in A has a subsequence converging to a point of A and A is hence compact.

To establish (7.4), we use the "**Cantor diagonalization**" method.[1]

Since A is strongly bounded, for each i there is an $M(i)$ such that $|f(z)| \leq M(i)$ for all z in $\operatorname{cl} U(z_i, r_i)$ and all f in A. Thus

$$\left| f_k^{(n)}(z_i) \right| \leq \frac{M(i)}{r_i^n} n!.$$

Now

$$f \mapsto \lambda_i^n(f)$$

---

[1]This method is often used in analysis.

form a countable set of mappings. Renumber these mappings as

$$\{\mu_1, \mu_2, \ldots, \mu_m, \ldots\}.$$

For $m = 1$,

$$\{\mu_1(f_k)\}_{k \in \mathbb{Z}_{>0}}$$

is a bounded sequence of complex numbers, and therefore there exists a subsequence B_1 of $\mathbb{Z}_{>0}$ such that

$$\lim_{\substack{k \in B_1 \\ k \to +\infty}} \mu_1(f_k) \text{ exists.}$$

For $m = 2$,

$$\{\mu_2(f_k)\}_{k \in B_1} \subseteq \{\mu_2(f_k)\}_{k \in \mathbb{Z}_{>0}}$$

is a bounded sequence of complex numbers. Therefore, there exists a subsequence B_2 of B_1 such that

$$\lim_{\substack{k \in B_2 \\ k \to +\infty}} \mu_2(f_k) \text{ exists.}$$

Continue to obtain a nested sequence of sets

$$B_1 \supseteq B_2 \supseteq \cdots \supseteq B_m \supseteq \cdots$$

with the property that

$$\lim_{\substack{k \in B_m \\ k \to +\infty}} \mu_m(f_k) \text{ exists.}$$

At last, *diagonalize* (justifying the name of the procedure); that is, let

$$B = \{n_1, \ n_2, \ \cdots, \ n_m, \cdots\},$$

where n_m is the m-th term of B_m. Then

$$\lim_{\substack{k \in B \\ k \to +\infty}} \mu_m(f_k) \text{ exists for all } m \in \mathbb{Z}_{>0}$$

because

$$\{n_m, \ n_{m+1}, \ \ldots\} \subseteq B_m.$$

$\square$

COROLLARY 7.29. *A set $A \subset \mathbf{H}(D)$ is compact if and only if it is closed and bounded in the strong sense.*

DEFINITION 7.30. *A set $A \subset \mathbf{H}(D)$ is* relatively compact *if* cl A *is compact. This definition clearly makes sense in much more general settings.*

COROLLARY 7.31. *Every strongly bounded subset of $\mathbf{H}(D)$ is relatively compact.*

DEFINITION 7.32. Let A be a bounded set in $\mathbf{H}(D)$, and let $\{f_k\}_{k=1}^{\infty}$ be a subset of A. We say that $f \in \mathbf{H}(D)$ is *adherent* to $\{f_k\}_{k=1}^{\infty}$ if it is a limit point of this sequence.

REMARK 7.33. If $\{x_k\}_{k=1}^{\infty}$ is a sequence in a compact topological space X and if every subsequence of $\{x_k\}_{k=1}^{\infty}$ that converges has the same limit, then the entire sequence $\{x_k\}_{k=1}^{\infty}$ converges.

THEOREM 7.34 (**Vitali's theorem**). *Let D be a domain in $\mathbb{C}$. Let $\{f_k\}_{k=1}^{\infty} \subset \mathbf{H}(D)$, and assume that the elements in this sequence are uniformly bounded on compact subsets of D. Let $S \subset D$, and assume that S has a limit point in D. Assume that $\lim_{k\to\infty} f_k(z)$ exists (pointwise) for all $z \in S$.*

Then the sequence $\{f_k\}$ converges uniformly on compact subsets of D.

PROOF. Every subsequence of the sequence $\{f_k\}$ has a converging sub-subsequence. Thus, there exists an f adherent to this sequence. Say that f and g are both adherent to $\{f_k\}$. Then

$$f(z) = \lim_{k\to\infty} f_k(z) = g(z) \quad \text{for} \quad z \in S$$

and thus, $f = g$. □

7.5. Approximation theorems and Runge's theorem

In this part of the chapter we consider the problem of approximating holomorphic functions by rational functions. We regard a nonconstant polynomial as a rational function whose only pole is at infinity and viceversa. The ability to uniformly approximate a holomorphic function depends on the region where the function is being approximated, as well as on the function. The strongest statement about uniform approximation of holomorphic functions that we will prove is known as Runge's theorem.

We observe first that we have already proved a form of Runge's theorem for an open disk Ω. A holomorphic function on Ω has a power series expansion at the center of the disk. For every positive integer n, we obtain a polynomial of degree n by discarding all the higher order terms in the series. These polynomials converge to the function uniformly on any compact subset of the disk.

On the other hand, we also know that uniform polynomial approximation does not hold in general. For instance, consider the punctured disk $\Omega = \{z \in \mathbb{C}; 0 < |z - z_0| < R\}$, with $R > 0$ and z_0 arbitrary, and the analytic function on Ω defined by $f(z) = \dfrac{1}{z - z_0}$ (we take advantage of the fact that it is a rational function whose only pole is at z_0 — not in Ω, of course); if f were uniformly approximated by a sequence of polynomials $\{p_n\}$ in the closed annulus $K = \{0 < r \le |z - z_0| \le \rho < R\}$, then by taking $\gamma(t) = \dfrac{r + \rho}{2} \exp(2\pi \imath t)$ for $0 \le t \le 2\pi$, we would obtain the contradiction that

$$0 = \lim_{n \to \infty} \int_\gamma p_n(z)\, dz = \int_\gamma f(z)\, dz = 2\pi\imath .$$

However, truncation of the Laurent series expansion for f on Ω shows that f is indeed uniformly approximated on K by rational functions whose poles lie outside Ω. This fact is generalized to arbitrary open sets Ω by Runge's theorem.

Specifically, in this section we prove that every holomorphic function defined on an open set can be uniformly approximated by rational functions whose poles lie in any given prescribed subset of $\widehat{\mathbb{C}}$, as long as the prescribed subset has nontrivial intersection with every component of the complement of the open set.

THEOREM 7.35 (Runge). *Let K be a compact subset of $\mathbb{C}$ and let S be a subset of $\widehat{\mathbb{C}} - K$ that intersects nontrivially each connected component of $\widehat{\mathbb{C}} - K$. If f is a holomorphic function on an open set $\Omega \supset K$, then for every $\epsilon > 0$, there exists a rational function R with poles only in S such that*

$$|f(z) - R(z)| < \epsilon \ \ \text{for all } z \in K.$$

The proof (based on S. A. Grabiner, *A short proof of Runge's theorem*, Amer. Math. Monthly, **83** (1976), 807-808) is given in Section 7.5.2.

Note that this is the implication 1) $\Longrightarrow$ 8) of the Fundamental Theorem. The converse follows from Theorem 7.2.

We begin with some preliminaries from real analysis needed in the proof of Runge's theorem. The proof depends on three major lemmas that are stated and proved in the subsequent sections; the first two are given in Section 7.5.3 and the third in 7.5.4. Runge's theorem follows from these lemmas.

Note that for $\widehat{\mathbb{C}} - K$ connected and $S = \{\infty\}$, Runge's theorem asserts that each function analytic in an open neighborhood of K can be approximated uniformly in K by a sequence of polynomials.

7.5.1. Preliminaries for the proof of Runge's theorem. We recall some terminology and notation. If F and K are subsets of $\mathbb{C}$

with F closed and K compact, then the distance between these two sets is the nonnegative real number

$$d(F, K) = \inf\{|z - w| \, ; z \in F \text{ and } w \in K\}$$

that satisfies

$$d(F, K) = 0 \text{ if and only if } F \cap K \neq \emptyset.$$

In particular, if $F = \{z_0\}$ is a set consisting of only one complex number, we set

$$d(z_0, K) = d(F, K) = \inf\{|z_0 - w| \, ; w \in K\}.$$

It is obvious that

$$d(z_0, K) = 0 \text{ if and only if } z_0 \in K.$$

If A and B are connected subsets of $\mathbb{C}$ that are not disjoint, then $A \cup B$ is connected. If C is a connected component of the set A, then C is an open subset of A.

We recall that if $\gamma : [a, b] \to \mathbb{C}$ is a curve, then its image is denoted range γ (see Section 4.1).

Next, we let K and S be as in the hypothesis of Runge's theorem, 7.35, and define $B(S)$ to be the set of continuous functions on K that are uniform limits of sequences of rational functions with poles only in S. The sums and products of elements of $B(S)$ are elements of $B(S)$, as well as the products of constants by elements of $B(S)$. The same holds for uniform limits of sequences in $B(S)$. We summarize these facts in a lemma. The proof is left as an exercise.

LEMMA 7.36. $B(S)$ *is an algebra that contains all rational functions with poles in S, and is closed under uniform limits in K.*

We will also need the following topological result.

LEMMA 7.37. *Let U and V be open subsets of $\mathbb{C}$ with $V \subseteq U$ and $\partial V \cap U = \emptyset$. If H is a connected component of U and $H \cap V \neq \emptyset$, then $H \subseteq V$.*

PROOF. Let $a \in H \cap V$, and let G be the connected component of V containing a. It is enough to show that $G = H$. But $H \cup G$ is connected, contained in U, and contains a. Since H is the component of U containing a, $G \subseteq H$. Furthermore, $\partial G \subseteq \partial V$ and so $\partial G \cap H = \emptyset$. This implies that $H - G = H - \operatorname{cl} G$ and, therefore, that $H - G$ is open in H. Since G is also open in H, the conclusion follows.   □

7.5.2. Proof of Runge's theorem. In this section we outline the proof of Runge's theorem. The details needed to fill in the outline will be completed in the next two sections.

(Outline). The proof consists of four steps:

(1) There exists a finite collection of oriented line segments γ_1, γ_2, ..., γ_n in $\Omega - K$ such that

$$f(z) = \frac{1}{2\pi\imath} \sum_{j=1}^{n} \int_{\gamma_j} \frac{f(\zeta)}{\zeta - z} \, d\zeta \text{ for all } z \in K.$$

This is the content of Lemma 7.38 of the next section.

(2) It suffices to prove that each integral $\int_{\gamma_j} \frac{f(\zeta)}{\zeta - z} d\zeta$ can be approximated uniformly on K by finite sums of rational functions $z \mapsto P\left(\frac{1}{z-c}\right)$, where P is a polynomial and $c \in S$. We hence drop the subscript j from the notation. We note that it will be convenient to regard $P\left(\frac{1}{z-\infty}\right)$ as a polynomial in z.

(3) We next note that the line integral $\int_{\gamma} \frac{f(\zeta)}{\zeta - z} d\zeta$ can be approximated uniformly on K by Riemann sums of the form

$$\sum_{k} \frac{a_k}{b_k - z}, \text{ with } a_k \in \mathbb{C} \text{ and } b_k \in \text{range } \gamma.$$

This is the content of Lemma 7.39.

(4) Finally, it suffices to show that each summand $\frac{a_k}{b_k - z}$ can be approximated uniformly on K by appropriate finite sums. This is Lemma 7.40 of Section 7.5.4.

$\square$

7.5.3. Two major lemmas. We prove two lemmas: The first gives an extension of the Cauchy Integral Formula, and the second provides an approximation by rational functions whose singularities lie on the curves over which we are integrating, for a function defined by the Cauchy Integral Formula.

LEMMA 7.38. *Let K be a compact subset of $\mathbb{C}$. If f is a holomorphic function on an open set $\Omega \supset K$, then there exists a finite collection of oriented line segments γ_1, γ_2, ..., γ_n in $\Omega - K$ such that*

$$f(z) = \frac{1}{2\pi\imath} \sum_{i=1}^{n} \int_{\gamma_i} \frac{f(\zeta)}{\zeta - z} d\zeta \quad \text{for all } z \in K. \tag{7.5}$$

PROOF. After enlarging K if necessary, we may assume that $K = \text{cl}(\text{int}(K))$. For example, we enlarge K if it consists of a single point.

Next, to simplify statements we adopt the standard convention that a curve $\gamma : [a, b] \to \widehat{\mathbb{C}}$, that is, the continuous function from a closed interval $[a, b]$ to the complex sphere, and its image range $\gamma = \{\gamma(t) \mid t \in [a, b] \}$ are both called γ. Thus by abuse of language, we say that a curve γ lies in a subset $T \subseteq \widehat{\mathbb{C}}$, when its image lies in T.

For any positive real number δ, we place a rectangular grid of horizontal and vertical lines in the plane $\mathbb{C}$ so that consecutive lines are at distance δ apart. We let $R_1, R_2, \ldots, R_m$ be the rectangles in the grid that have nonempty intersection with K. Since K is compact, there are only a finite number of such rectangles. We can (and from now on do) choose δ such that $R_j \subset \Omega$ for all j; if $\Omega = \mathbb{C}$ any $\delta > 0$ suffices, and otherwise, it is enough to consider any $0 < \delta < \frac{1}{2}d(K, \mathbb{C}-\Omega)$, since $z \in R_j$ implies that $d(z, K) < \sqrt{2}\delta$.

We denote the boundary of R_j by ∂R_j and orient it to be traversed in the counterclockwise direction. We observe that integration of a continuous form along the common boundaries of any pair of contiguous R_i and R_j cancel out (as in the proof of Goursat's theorem 4.52).

This last observation implies that we can choose a set of curves $\mathcal{S}$ whose images are a subset of the sides in $\cup_{j=1}^m \partial R_j$, and such that the set $\mathcal{S} = \{\gamma_i; 1 \leq i \leq n\}$ satisfies

(1) if γ_i is in $\mathcal{S}$, it lies on a side of only one R_j,
(2) if γ_i is in $\mathcal{S}$, then it is disjoint from K, and
(3) for any continuous function g on $\cup_{j=1}^m \partial R_j$, we have

$$\sum_{j=1}^m \int_{\partial R_j} g = \sum_{i=1}^n \int_{\gamma_i} g. \tag{7.6}$$

Note that we have identified the curves with their images. Since the boundary of each rectangle is oriented, each γ_i is an oriented line segment in $\Omega - K$. We now prove that these γ_i also satisfy equation (7.5), the equation involving the function f given in the statement of the lemma.

If $z \in K$ and z is not on the boundary of any of the rectangles, then the function

$$w \mapsto g(w) = \frac{1}{2\pi\imath} \cdot \frac{f(w)}{z - w}, \quad w \in \cup_{j=1}^m \partial R_j$$

is continuous. Thus we have by (7.6)

$$\frac{1}{2\pi\imath} \sum_{j=1}^m \int_{\partial R_j} \frac{f(w)}{w - z} \, dw = \frac{1}{2\pi\imath} \sum_{i=1}^n \int_{\gamma_i} \frac{f(w)}{w - z} \, dw.$$

Now if z belongs to the interior of exactly one of the R_j, call it R_t. If $j \neq t$, then $z \notin R_j$ and

$$\frac{1}{2\pi\iota} \int_{\partial R_j} \frac{f(w)}{w - z}\, dw = 0;$$

also, since $z \in R_t$, the Cauchy Integral formula says

$$\frac{1}{2\pi\iota} \int_{\partial R_t} \frac{f(w)}{w - z}\, dw = f(z)\,.$$

Thus

$$f(z) = \frac{1}{2\pi\iota} \sum_{i=1}^{n} \int_{\gamma_i} \frac{f(w)}{w - z}\, dw\,. \tag{7.7}$$

Since range γ_i does not intersect K, both sides of equation (7.7) are continuous functions on K and they agree on the set of z in K that are not on the boundary of any rectangle, a dense subset of K. Thus they agree for all $z \in K$. □

LEMMA 7.39. *Let γ be any pdp, and let K be a compact set not meeting the image of γ. If f is continuous on γ and ϵ is any positive real number, then there is a rational function R having all of its poles on* range γ *such that*

$$\left| \int_\gamma \frac{f(w)}{z - w}\, dw - R(z) \right| < \epsilon \quad \text{for all } z \in K.$$

PROOF. Since K and the image of γ are disjoint, $d(K, \text{range}\,\gamma) > 0$, and we can choose a number r with $0 < r < d(K, \text{range}\,\gamma)$.

Assuming that γ is parametrized by $[0, 1]$, then for all $0 \leq s, t \leq 1$ and all $z \in K$, we have

$$\left| \frac{f(\gamma(t))}{\gamma(t) - z} - \frac{f(\gamma(s))}{\gamma(s) - z} \right|$$

$$\leq \frac{1}{r^2} |f(\gamma(t))\gamma(s) - f(\gamma(s))\gamma(t) - z\,(f(\gamma(t)) - f(\gamma(s)))|$$

$$\leq \frac{1}{r^2} |f(\gamma(t))| \cdot |\gamma(s) - \gamma(t)| + \frac{1}{r^2} |\gamma(t)| \cdot |f(\gamma(s)) - f(\gamma(t))| +$$

$$\frac{|z|}{r^2} |f(\gamma(s)) - f(\gamma(t))|\,.$$

Since γ and f are continuous functions and K is a compact set, there is a constant $C > 0$ such that $|z| < \frac{C}{2}$ for all $z \in K$, $|\gamma(t)| \leq \frac{C}{2}$ and

$|f(\gamma(t))| \leq C$ for all $t \in [0,1]$. Thus for all s and t in $[0,1]$ and all $z \in K$,

$$\left| \frac{f(\gamma(t))}{\gamma(t) - z} - \frac{f(\gamma(s))}{\gamma(s) - z} \right| \leq \frac{C}{r^2} [\, |\gamma(s) - \gamma(t)| + |f(\gamma(s)) - f(\gamma(t))| \,].$$

Since both γ and $f \circ \gamma$ are uniformly continuous on $[0,1]$, there is a partition of $[0,1]$ with $0 = t_0 < t_1 < \cdots t_n = 1$ such that

$$\left| \frac{f(\gamma(t))}{\gamma(t) - z} - \frac{f(\gamma(t_j))}{\gamma(t_j) - z} \right| < \frac{\epsilon}{L(\gamma)} \tag{7.8}$$

for $t_{j-1} \leq t \leq t_j$, $1 \leq j \leq n$, and all $z \in K$, where $L(\gamma)$ denotes the length of γ (recall Definition 4.53).

Define the function R as follows. For $z \neq \gamma(t_j)$,

$$R(z) = \sum_{j=1}^{n} f(\gamma(t_j)) \cdot \frac{\gamma(t_j) - \gamma(t_{j-1})}{\gamma(t_j) - z}.$$

Then R is a rational function whose poles are contained in the set

$$\{\gamma(t_1), \gamma(t_2), \ldots, \gamma(1)\};$$

in particular, they are contained in range γ. Now equation (7.8) gives

$$\left| \int_{\gamma} \frac{f(w)}{w - z} \, dw - R(z) \right| = \left| \sum_{j=1}^{n} \int_{t_{j-1}}^{t_j} \left(\frac{f(\gamma(t))}{\gamma(t) - z} - \frac{f(\gamma(t_j))}{\gamma(t_j) - z} \right) \gamma'(t) \right| dt$$

$$\leq \frac{\epsilon}{L(\gamma)} \sum_{j=1}^{n} \int_{t_{j-1}}^{t_j} |\gamma'(t)| \, dt = \epsilon \quad \text{for all } z \in K.$$

$\square$

7.5.4. Approximating $\dfrac{1}{z - a}$. Of central importance in the proof of Runge's theorem is

LEMMA 7.40. *If $a \in \mathbb{C} - K$, then the function defined by $z \mapsto (z - a)^{-1} \in B(S)$.*

PROOF. If $\infty \in S$, then for any z_0 in the unbounded component of $\mathbb{C} - K$ such that $|z_0|$ is sufficiently large, the Taylor series for the rational function $g : z \mapsto \dfrac{1}{z - z_0}$ converges uniformly on K and therefore g is in $B(S)$.

We claim that $B((S - \{\infty\}) \cup \{z_0\}) \subseteq B(S)$. Indeed, if $f \in B((S - \{\infty\}) \cup \{z_0\})$ and R is a rational function with poles in $(S - \{\infty\}) \cup \{z_0\}$ uniformly approximating f on K, we can write $R = R_1 + R_2$, where R_1 has all its poles (if any) in $S - \{\infty\}$ and R_2 has a unique pole (if any) at z_0. But then R_2 can be approximated uniformly by polynomials P_2 on K, and therefore, $R_1 + P_2$ have poles only in S and approximate uniformly f on K. That is, $f \in B(S)$.

Thus it is sufficient to prove the lemma for $S \subset \mathbb{C}$. We will use Lemma 7.37.

Let $U = \mathbb{C} - K$, and let $V = \{w \in \mathbb{C}; z \mapsto (z - w)^{-1} \in B(S)\}$. Then $S \subseteq V \subseteq U$. We want to show $U = V$. We will show first that

$$\text{if } a \in V \text{ and } |b - a| < d(a, K), \text{ then } b \in V. \tag{7.9}$$

Assume $a \in V$ and $|b - a| < d(a, K)$. Then there is a real number r, $0 < r < 1$, such that $|b - a| < r\,|z - a|$ for all $z \in K$. Note that

$$(z - b)^{-1} = (z - a)^{-1} \left(1 - \frac{b - a}{z - a}\right)^{-1}; \tag{7.10}$$

since $\dfrac{|b - a|}{|z - a|} < r < 1$ for all $z \in K$, we can use the Weierstrass M-test to conclude that the series (in the variable z)

$$\left(1 - \frac{b - a}{z - a}\right)^{-1} = \sum_{n=0}^{\infty} \left(\frac{b - a}{z - a}\right)^n \tag{7.11}$$

converges uniformly on K. Now Lemma 7.36 and equation (7.10) imply that (7.9) holds.

Note that (7.9) says that V is an open subset of $\mathbb{C}$.

Next we show that $\partial V \cap U = \emptyset$.

If $b \in \partial V$, let $\{a_n\}$ be a sequence of elements of V converging to b. Since $b \notin V$, it follows that $|b - a_n| \geq d(a_n, K)$, and letting $n \to \infty$, we obtain $0 = d(b, K)$; that is, $b \in K$ and therefore $b \notin U$.

We now apply Lemma 7.37. If H is any connected component of $U = \mathbb{C} - K$, then by the definition of S, there exists $s \in H \cap S$. But then $s \in H \cap V \neq \emptyset$ and the lemma implies $H \subseteq V$. Therefore every connected component of U lies in V and consequently $U \subseteq V$, and thus $U = V$. □

Proof of Runge's theorem. Lemmas 7.38, 7.39, and 7.40, respectively, complete steps 1, 3, and 4 of the proof. □

COROLLARY 7.41. *If D is any simply connected domain in the plane and f is a holomorphic function in D, then f can be approximated uniformly on all compact subsets of D by polynomials.*

Exercises

7.1. Show that Theorem 7.2 has no analogue for real variables in that the absolute value function on $\mathbb{R}$, which has no derivative at 0, can be approximated uniformly by differentiable functions.

7.2. Construct an example of a sequence of real differentiable functions converging uniformly to a real differentiable function on a closed interval such that the sequence given by the derivatives does not converge uniformly there.

(Hint: The sequence $f_n(x) = x^n$ does not converge uniformly on $[0, 1]$.)

7.3. Show that the series $\displaystyle\sum_{n=-\infty}^{\infty} \frac{1}{z - n}$ does not converge.

7.4. Prove Lemma 7.25.

7.5. Prove Lemma 7.36: that $B(S)$ is an algebra closed under uniform limits on K.

CHAPTER 8

Conformal Equivalence

In this chapter we study *conformal maps* between domains in the extended complex plane. These maps are one-to-one meromorphic functions. Our goal is characterize all simply connected domains in the complex plane. The first two sections of this chapter study the action of a quotient of the group of two-by-two nonsingular complex matrices on the extended complex plane $\widehat{\mathbb{C}}$, namely the group $\mathrm{PSL}(2, \mathbb{C})$, the projective special linear group. This group is also known as the Möbius group. In the third section we characterize simply connected proper domains in the complex plane by establishing the Riemann Mapping Theorem. This extraordinary theorem tells us that there are conformal maps between any two such domains. The study of the Möbius group is connected intimately with hyperbolic geometry. In the next-to-last section of this chapter, we define the non-Euclidean metric, also known as the hyperbolic metric and the Poincaré metric. Hyperbolic geometry has increasingly become an essential part of complex variable theory. We end the chapter by using the Schwarz lemma to establish the deep connection between complex variables and geometry, Theorem 8.43, which says that a holomorphic mapping is either an isometry or a contraction in the hyperbolic metric. The last section is devoted to a study of certain bounded analytic functions on the unit disk known as *finite Blaschke products*.

We begin with

DEFINITION 8.1. A one-to-one meromorphic function is called a *conformal map*.

This is the correct notion of isomorphism in the category of meromorphic mappings, since the inverse of a conformal map is also conformal.[1] Thus the concept introduces a natural equivalence relation on the family of domains on the sphere, called *conformal equivalence*.

---

[1] In geometry, C^1-maps are called *conformal* if they preserve angles. We have seen in Proposition 6.21 of Chapter 5 that in the orientation preserving case, these are precisely the holomorphic functions with nowhere vanishing derivatives. Thus the two definitions agree locally for sense preserving transformations. In our definition, we also require injectivity.

DEFINITION 8.2. Let D be a domain in $\widehat{\mathbb{C}}$. Aut D is defined as the group (under composition) of conformal automorphisms of D; that is, it consists of the conformal maps from D onto itself.

There are two naturally related problems:

Problem I. Describe Aut D for a given D.
Problem II. Given two domains D and D', determine when they are conformally equivalent.

We solve Problem I for $D = \widehat{\mathbb{C}}$, $D = \mathbb{C}$, and $D = \mathbb{D}$ (the unit disk $\{z \in \mathbb{C};\ |z| < 1\}$ in $\widehat{\mathbb{C}}$), and Problem II for D and D' any pair of simply connected domains in $\widehat{\mathbb{C}}$.

8.1. Fractional linear (Möbius) transformations

In this section we describe the (orientation preserving) Möbius group, and show that for the domains $D = \widehat{\mathbb{C}}$, $\mathbb{C}$, a disk or a half-plane, the group Aut D is a subgroup of this group.

DEFINITION 8.3. A *fractional linear transformation* (or *Möbius transformation*) is a map $A : \widehat{\mathbb{C}} \to \widehat{\mathbb{C}}$ of the form

$$z \mapsto A(z) = \begin{cases} \dfrac{az+b}{cz+d} & \text{if } c \neq 0 \text{ and } z \neq \infty \text{ and } z \neq -\frac{d}{c}, \\ \frac{a}{c} & \text{if } c \neq 0 \text{ and } z = \infty, \\ \infty & \text{if } c \neq 0 \text{ and } z = -\frac{d}{c}, \\ \frac{a}{d}z + \frac{b}{d} & \text{if } c = 0 \text{ and } z \neq \infty, \\ \infty & \text{if } c = 0 \text{ and } z = \infty, \end{cases} \tag{8.1}$$

where a, b, c, and d are complex numbers such that $ad - bc \neq 0$.

Without loss of generality we assume from now on that $ad - bc = 1$ (*the reader should prove* that there is really no loss of generality in this assumption; that is, establish Exercise 8.1). Also, whenever convenient we will multiply each of the four constants a, b, c, and d by -1, since this does not alter the Möbius transformation nor the condition $ad - bc = 1$. We abbreviate

$$A(z) = \frac{az+b}{cz+d},$$

since all special values in (8.1) are obtained as limits of expressions of this form with both z and $\frac{az+b}{cz+d}$ finite.

REMARK 8.4. A Möbius transformation is an element of $\mathrm{Aut}(\widehat{\mathbb{C}})$, and the set of all Möbius transformations form a group under composition, the *Möbius group*.

REMARK 8.5. Other related groups are the matrix group

$$\mathrm{SL}(2,\mathbb{C}) = \left\{ \begin{bmatrix} a & b \\ c & d \end{bmatrix} ; \ a,b,c,d \in \mathbb{C}, ad - bc = 1 \right\},$$

the corresponding quotient group

$$\mathrm{PSL}(2,\mathbb{C}) = \mathrm{SL}(2,\mathbb{C})/\{\pm I\},$$

where $I = \begin{bmatrix} 1 & 0 \\ 0 & 1 \end{bmatrix}$ is the identity matrix, and the *extended Möbius group* of orientation preserving and reversing transformations, consisting of the motions

$$z \mapsto \frac{az+b}{cz+d} \quad \text{and} \quad z \mapsto \frac{a\bar{z}+b}{c\bar{z}+d}, \quad \text{with } ad - bc = 1.$$

Here orientation reversing means that angles are preserved in magnitude but reversed in sense (as the map $z \to \bar{z}$ does).

It is clear that

$$1 \to \{\pm I\} \to \mathrm{SL}(2,\mathbb{C}) \to \mathrm{Aut}(\widehat{\mathbb{C}}) \tag{8.2}$$

is an exact sequence, where the first two arrows denote inclusion and by the last arrow a matrix $\begin{bmatrix} a & b \\ c & d \end{bmatrix}$ in $\mathrm{SL}(2,\mathbb{C})$ is sent to the element of $\mathrm{Aut}(\widehat{\mathbb{C}})$ given by (8.1). That is, for any pair of consecutive maps in the sequence, the kernel of the second map coincides with the image of the first one.

It is also clear that the image of the last arrow in the sequence (8.2) is precisely the Möbius group and, therefore, that it is isomorphic to $\mathrm{PSL}(2,\mathbb{C})$, the quotient of $\mathrm{SL}(2,\mathbb{C})$ by $\pm I$ as defined above.

It is natural to ask whether the last arrow is surjective; that is, whether the Möbius group coincides with $\mathrm{Aut}(\widehat{\mathbb{C}})$. We will see that this is the case in Theorem 8.17.

Let A be an element of $\mathrm{PSL}(2,\mathbb{C})$. Take the trace of a preimage of A in $\mathrm{SL}(2,\mathbb{C})$, and square it. This quantity will be the same for either of the two preimages of A. Thus even though the trace of an element in the Möbius group is not well defined, *the trace squared* of an element in $\mathrm{PSL}(2,\mathbb{C})$ is well defined.

DEFINITION 8.6. For A in the Möbius group, described by (8.1) with $ad - bc = 1$, we define $\operatorname{tr}^2 A = (a + d)^2$ where A is the equivalence class of the matrices $\pm \begin{bmatrix} a & b \\ c & d \end{bmatrix}$ in $\operatorname{SL}(2, \mathbb{C})$.

8.1.1. Fixed points of Möbius transformations. Let A be any element of the Möbius group different from the identity map. We are interested in the *fixed points* of A; that is, those $z \in \widehat{\mathbb{C}}$ with $A(z) = z$. Thus if $A(z) = \dfrac{az + b}{cz + d}$ with (the standard normalization) $ad - bc = 1$, then for a fixed point z of A we have either $z = \infty$, or $z \in \mathbb{C}$ and $cz^2 + (d - a)z + b = 0$. We consider two cases:

Case 1: $c = 0$. In this case ∞ is a fixed point of A and $ad = 1$. If $a = d$, then $A(z) = z + b$ with $b \neq 0$ (because A is not the identity map), and A has no other fixed point. If $a \neq d$, then A has one more fixed point, given by $\zeta = \dfrac{b}{d - a}$.

We note that in this case A has precisely one fixed point if and only if $A(z) = z + b$ with $b \in \mathbb{C}$ and $b \neq 0$.

Case 2: $c \neq 0$. In this case, ∞ is not fixed by A, and its fixed points are given by

$$\frac{a - d \pm \sqrt{(a - d)^2 + 4bc}}{2c} = \frac{(a - d) \pm \sqrt{\operatorname{tr}^2 A - 4}}{2c}.$$

We have thus proved the following result.

PROPOSITION 8.7. *If $A(z) = \dfrac{az + b}{cz + d}$ with $ad - bc = 1$ is a Möbius transformation different from the identity map, then A has either one or two fixed points in $\widehat{\mathbb{C}}$. It has exactly one if and only if $\operatorname{tr}^2 A = 4$.*

8.1.2. cross-ratios.

PROPOSITION 8.8. *Given three distinct points z_2, z_3, z_4 in $\widehat{\mathbb{C}}$, there exists a unique Möbius transformation S with $S(z_2) = 1$, $S(z_3) = 0$, and $S(z_4) = \infty$.*

PROOF. The proof has two parts.

Uniqueness: If S_1 and S_2 are Möbius transformations that solve our problem, then $S_1 \circ S_2^{-1}$ is a Möbius transformation that fixes 1, 0, and ∞, and hence, by Proposition 8.7, it is the identity map.

Existence: If the z_i are complex numbers, then

$$S(z) = \frac{z - z_3}{z - z_4} \cdot \frac{z_2 - z_4}{z_2 - z_3}.$$

is the required map.

If one of the $z_i = \infty$, use a limiting procedure to obtain the following.

If $z_2 = \infty$,   then $S(z) = \dfrac{z - z_3}{z - z_4}$,

If $z_3 = \infty$,   then $S(z) = \dfrac{z_2 - z_4}{z - z_4}$,

If $z_4 = \infty$,   then $S(z) = \dfrac{z - z_3}{z_2 - z_3}$.

$\square$

COROLLARY 8.9. *If z_i and w_i (i = 2, 3, 4) are two triples of distinct points in $\widehat{\mathbb{C}}$, then there exists a unique Möbius transformation S such that $S(z_i) = w_i$; thus the Möbius group is uniquely triply transitive on $\widehat{\mathbb{C}}$.*

DEFINITION 8.10. The *cross-ratio* (z_1, z_2, z_3, z_4) of four distinct points in $\widehat{\mathbb{C}}$ is the image of z_1 under the Möbius transformation taking z_2 to 1, z_3 to 0, and z_4 to ∞; that is,

$$(z_1, z_2, z_3, z_4) = \frac{z_1 - z_3}{z_1 - z_4} \cdot \frac{z_2 - z_4}{z_2 - z_3}$$

if the four points are finite, with the corresponding limiting values if one of the z_i equals ∞.

PROPOSITION 8.11. *If z_1, z_2, z_3, and z_4 are four distinct points in $\widehat{\mathbb{C}}$ and T is any Möbius transformation, then*

$$(T(z_1), T(z_2), T(z_3), T(z_4)) = (z_1, z_2, z_3, z_4).$$

PROOF. If we define $S(z) = (z, z_2, z_3, z_4)$ for $z \in \widehat{\mathbb{C}}$, then $S \circ T^{-1}$ is a Möbius transformation taking $T(z_2)$ to 1, $T(z_3)$ to 0, and $T(z_4)$ to ∞. Therefore

$$(T(z_1), T(z_2), T(z_3), T(z_4)) = S \circ T^{-1}(T(z_1)) = S(z_1).$$

$\square$

DEFINITION 8.12. A *circle* in $\widehat{\mathbb{C}}$ is either an Euclidean (ordinary) circle in $\mathbb{C}$ or a straight line in $\mathbb{C}$ together with ∞ (this is a circle passing through ∞). See Exercise 3.19.

PROPOSITION 8.13. *The cross-ratio (z_1, z_2, z_3, z_4) is a real number if and only if the four points lie on a circle in $\widehat{\mathbb{C}}$.*

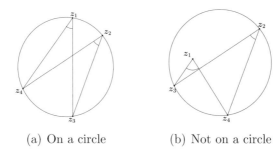

(a) On a circle   (b) Not on a circle

FIGURE 8.1. The cross-ratio arguments

PROOF. This is an elementary geometric argument that goes as follows. It is clear that

$$\arg(z_1, z_2, z_3, z_4) = \arg \frac{z_1 - z_3}{z_1 - z_4} - \arg \frac{z_2 - z_3}{z_2 - z_4}.$$

It is also clear from the geometry of the situation (see Figure 8.1 and Exercise 8.3) that the two quantities on the right hand differ by πn, with $n \in \mathbb{Z}$, if and only if the four points lie on a circle in $\widehat{\mathbb{C}}$. □

THEOREM 8.14. *A Möbius transformation maps circles to circles.*

PROOF. This follows immediately from Propositions 8.11 and 8.13. □

We use the following *standard notation* in the rest of this chapter: $\mathbb{D}$ denotes the unit disk $\{z \in \mathbb{C}; |z| < 1\}$, and $\mathbb{H}^2$ the upper half-plane $\{z \in \mathbb{C}; \Im z > 0\}$. Note that both $\mathbb{D}$ and $\mathbb{H}^2$ should be regarded as disks in $\widehat{\mathbb{C}}$, since they are bounded by circles in $\widehat{\mathbb{C}}$: the unit circle S^1 and the extended real line $\widehat{\mathbb{R}} = \mathbb{R} \cup \{\infty\}$, respectively.

COROLLARY 8.15. *If* $w(z) = \dfrac{z - \imath}{z + \imath}$ *for* $z \in \mathbb{H}^2$, *then* w *is a conformal map of* $\mathbb{H}^2$ *onto* $\mathbb{D}$.

PROOF. w maps $\widehat{\mathbb{R}} = \mathbb{R} \cup \{\infty\}$ onto S^1 (the unit circle), $w(\imath) = 0$, and Möbius transformations are conformal. □

8.2. Aut(D) for $D = \widehat{\mathbb{C}}$, $\mathbb{C}$, $\mathbb{D}$, and $\mathbb{H}^2$

THEOREM 8.16. *An entire function f belongs to* Aut$(\mathbb{C})$ *if and only if there exist a and b in $\mathbb{C}$, $a \neq 0$, with $f(z) = az + b$ for all $z \in \widehat{\mathbb{C}}$.*

PROOF. The if part is trivial.

For the only if part, since f is entire, we can write

$$f(z) = \sum_{n=0}^{\infty} a_n z^n \quad \text{for all } z \in \mathbb{C} .$$

If ∞ were an essential singularity of f, then $f(|z| > 1)$ would be dense in $\mathbb{C}$. But

$$f(|z| > 1) \cap f(|z| < 1) \text{ is empty.}$$

Thus there is a nonnegative integer N such that $a_n = 0$ for all $n > N$ and such that $a_N \neq 0$; that is, f is a polynomial of degree N. If N were bigger than one or equal to zero, then f would not be injective. □

THEOREM 8.17. $\text{Aut}(\widehat{\mathbb{C}}) \cong \text{PSL}(2, \mathbb{C})$. *Thus the last arrow in the exact sequence (8.2) corresponds to a surjective map.*

PROOF. We need only show that $\text{Aut}(\widehat{\mathbb{C}}) \subset \text{PSL}(2, \mathbb{C})$. Let f be an element of $\text{Aut}(\widehat{\mathbb{C}})$. If $f(\infty) = \infty$, then f is a Möbius transformation, by Theorem 8.16. If $f(\infty) = \zeta \neq \infty$, then consider $A(z) = \dfrac{1}{z - \zeta}$ and conclude that $A \circ f$ fixes ∞. □

THEOREM 8.18. $A \in \text{Aut}(\mathbb{D})$ *if and only if there exist a and b in $\mathbb{C}$ such that* $|a|^2 - |b|^2 = 1$ *and* $A(z) = \dfrac{az + b}{\overline{b}z + \overline{a}}$ *for all $z \in \mathbb{D}$.*

PROOF. **if part:** Assume that A is of the above form, and observe that $a \neq 0$. We must show that $A \in \text{Aut}(\mathbb{D})$. This will follow from the following easy-to-prove facts:

(1) Mappings A of the given form constitute a group under composition.

(2) $|z| = 1$ if and only if $|A(z)| = 1$.

(3) $|A(0)| = \left|\dfrac{b}{a}\right| < 1$.

(4) $A(\mathbb{D})$ is connected. Thus either $A(\mathbb{D})$ is contained in $\mathbb{D}$ or $A(\mathbb{D}) \cap \mathbb{D}$ is empty. From (3) we see that $A(\mathbb{D}) \subseteq \mathbb{D}$.

(5) Obviously $A \circ A^{-1}(\mathbb{D}) = \mathbb{D}$, which implies that $A(\mathbb{D}) = \mathbb{D}$.

    only if part: Let $f \in \text{Aut}(\mathbb{D})$ and $w = f(z)$. Then $z = f^{-1}(w)$ and $f^{-1} \in \text{Aut}(\mathbb{D})$.

(6) If $f(0) = 0$, then

$$|f(z)| \le |z| \quad \text{and} \quad |z| = \left|f^{-1}(w)\right| \le |w| = |f(z)|.$$

Thus, by Schwarz's lemma, there exists a $\theta \in \mathbb{R}$ such that $f(z) = e^{i\theta} z$ for all $z \in \mathbb{D}$. So we can take $a = e^{i\frac{\theta}{2}}$ and $b = 0$ to conclude that f has the required form.

(7) If $f(0) = \zeta \ne 0$, then $0 < |\zeta| < 1$ and we set $A(z) = \dfrac{z - \zeta}{1 - \bar{\zeta} z}$.
The Möbius transformation A belongs to $\mathrm{Aut}(\mathbb{D})$, and $A \circ f$ fixes the origin.

$\square$

Just as in Section 8.1 we defined $\mathrm{PSL}(2, \mathbb{C})$ and then proved that it is isomorphic to the group $\mathrm{Aut}(\widehat{\mathbb{C}})$, we can define the group $\mathrm{PSL}(2, \mathbb{R}) = \mathrm{SL}(2, \mathbb{R})/\{\pm I\}$ of appropriate matrices with real coefficients modulo plus or minus the identity matrix and obtain the following description.

THEOREM 8.19. $\mathrm{Aut}(\mathbb{H}^2) \cong \mathrm{PSL}(2, \mathbb{R})$.

PROOF. This is an easy consequence of Corollary 8.15 and the preceding theorem. $\square$

8.3. The Riemann Mapping Theorem

We now combine the results about Möbius transformations of the previous two sections with results from Chapter 7 about compact and bounded families of holomorphic functions to show that every simply connected plane domain, other than $\mathbb{C}$ itself, is conformally equivalent to the unit disk; any conformal map from a domain D onto the unit disk $\mathbb{D}$ will be called a *Riemann Map*.

Recall that a set A is a *proper subset* of a set B if $A \subset B$ and $A \ne B$.

THEOREM 8.20 (**Riemann Mapping Theorem**). *Let D be a proper simply connected open subset of $\mathbb{C}$ and let $\zeta \in D$. Then there exists a unique conformal map $f : D \to \mathbb{D}$ with $f(\zeta) = 0$, $f'(\zeta) > 0$, and $f(D) = \mathbb{D}$.*

PROOF. The argument has two parts.

Existence. We first reduce the problem to a special case.

FIRST REDUCTION: It suffices to assume that D is bounded.

PROOF. Since $D \neq \mathbb{C}$, we can choose $b \in \mathbb{C} - D$. Since D is simply connected there is a branch $g(z)$ of $\log(z - b)$ on D. Thus

$$e^{g(z)} = z - b \text{ for all } z \in D.$$

The function g is injective: For if $g(w) = g(z)$, then $w - b = z - b$. Furthermore, if $c \in D$, then $g(z) - g(c) \neq 2\pi\imath$ for all $z \in D$. Otherwise

$$z - b = e^{g(z)} = e^{g(c)+2\pi\imath} = e^{g(c)} = c - b.$$

Choose $c \in D$ and $\delta > 0$ such that

$$|w - g(c)| < \delta \Rightarrow w \in g(D).$$

Thus

$$|w - g(c) - 2\pi\imath| < \delta \Rightarrow w \notin g(D).$$

Now

$$F(z) = \frac{1}{g(z) - g(c) - 2\pi\imath}$$

is an isomorphism of D onto $F(D)$, and the domain $F(D)$ is contained in $\operatorname{cl} U\left(0, \dfrac{1}{\delta}\right)$. $\square$

We are now reduced to solving the mapping problem for $(F(D), F(\zeta))$.

SECOND REDUCTION: We may also assume that $\zeta = 0$. Toward this end, set $\alpha = F(\zeta)$.

PROOF. Look at $G(z) = e^{-\imath\theta}(z - \alpha)$, where $\theta = \arg F'(\zeta)$. If $f : G(F(D)) \to \mathbb{D}$ is a conformal surjective map with $f(0) = 0$ and $f'(0) > 0$, then

$$h = f \circ G \circ F \text{ is a conformal map of } D \text{ onto } \mathbb{D}, \text{ with}$$

$$h(\zeta) = 0 \text{ and } h'(\zeta) = f'(0)G'(\alpha)F'(\zeta) > 0.$$

$\square$

Thus we now assume that D is bounded, simply connected, and $\zeta = 0 \in D$.

PROOF OF THE THEOREM UNDER THESE ASSUMPTIONS.

We define

$$\mathcal{F} = \{f \in \mathbf{H}(D); f \text{ is either conformal or identically zero,}$$

$$f(0) = 0, f'(0) \in \mathbb{R}, f'(0) \geq 0 \text{ and } |f(z)| < 1 \text{ all } z \in D\}.$$

Our first observation is that $\mathcal{F}$ is nonempty. Of course, $f \equiv 0$ is in $\mathcal{F}$. This is not good enough for much. Since D is bounded, there exists an $M > 0$ such that $|z| \leq M$ for all $z \in D$. Hence if $c \in \mathbb{R}$ and $c > M$, then $f(z) = \dfrac{z}{c}$ for $z \in D$ defines a function in $\mathcal{F}$ (which is not identically zero).

Next we show that $\mathcal{F}$ is compact. The elements of $\mathcal{F}$ are bounded by 1; hence, this family is certainly strongly bounded.

To show that it is closed, let $\{f_n\} \subset \mathcal{F}$ be such that $f_n \to f$ uniformly on all compact subsets of D. Then $f \in \mathbf{H}(D)$, and since each f_n vanishes at 0, so does f.

It is now convenient to consider two cases:

I. $f_n \equiv 0$ for infinitely many distinct n.
In this case, $f \equiv 0$ and hence certainly $f \in \mathcal{F}$.

II. $f_n \equiv 0$ for finitely many n.
In this case we may assume that each f_n is a conformal map so that $f_n'(0) > 0$, and thus, $f'(0) \geq 0$. Hurwitz's theorem 7.8 says that f is either constant (hence identically zero) or univalent. Since $|f_n(z)| < 1$ for all $z \in D$, we conclude that $|f(z)| \leq 1$ for all $z \in D$. If $|f(z_0)| = 1$ for some $z_0 \in D$, then $|f| \equiv 1$ by the Maximum Modulus Principle; this is a contradiction to $f(0) = 0$. Thus $|f(z)| < 1$ for all $z \in D$, and we conclude that $f \in \mathcal{F}$. Thus $\mathcal{F}$ is closed and therefore compact.

If $S = \{f'(0); f \in \mathcal{F}\}$, then $S \subset \mathbb{R}_{\geq 0}$. We claim that S is bounded from above. Indeed, choose $\epsilon > 0$ so that $\operatorname{cl} U(0, \epsilon) \subset D$. If $\gamma(\theta) = \epsilon\, e^{\imath\theta}$ for $0 \leq \theta \leq 2\pi$ is the circle centered at 0 with radius ϵ, then for any $f \in \mathcal{F}$, we have

$$f'(0) = \frac{1}{2\pi\imath} \int_\gamma \frac{f(\zeta)}{\zeta^2}\, d\zeta$$

and thus

$$|f'(0)| \leq \frac{1}{2\pi} \frac{2\pi\epsilon}{\epsilon^2} = \frac{1}{\epsilon}.$$

If $\mu = \sup S$, then $\mu \geq \dfrac{1}{M} > 0$ because $\dfrac{1}{c} \in S$ for all $c > M$, as we saw above. Also, there exists a sequence $\{f_n\} \subset \mathcal{F}$ such that

$$\lim_{n\to\infty} f_n'(0) = \mu\,.$$

Since $\mathcal{F}$ is compact, there exists a convergent subsequence $\{f_{n_k}\}$ with $\lim\limits_{k\to\infty} f_{n_k} = f \in \mathcal{F}$. Since $f'(0) = \mu$, f is a conformal map. Also $f(D) \subseteq \mathbb{D}$.

We need to show that $f(D) = \mathbb{D}$. We show that if we assume $f(D) \neq \mathbb{D}$, we can construct an $h \in \mathcal{F}$ with $h'(0) > \mu$ and thus contradict the fact that $\mu = \sup S$. Namely, if $f(D) \neq \mathbb{D}$, then there exists $w_0 = re^{i\theta}$ with $0 < r < 1$ such that $w_0 \in \partial f(D)$. We now construct h as follows:

(1) Let $g_1(z) = e^{-i\theta} f(z)$. The map g_1 is the map f followed by a rotation through the angle $-\theta$ and it sends w_0 to r.

(2) Let $p(z) = \dfrac{r - g_1(z)}{1 - rg_1(z)}$. The map p is g_1 followed by an automorphism of $\mathbb{D}$ that sends r to 0 (see Exercise 8.4).

   Note that $p(z) \neq 0$ for all $z \in D$ ($p(z) = 0$ if and only if $g_1(z) = r$ if and only if $f(z) = w_0$).

(3) Let $q(z) = p(z)^{\frac{1}{2}}$, where we choose the branch of the square root[2] with $q(0) = r^{\frac{1}{2}} > 0$.

   The map q is injective because $q(z_1) = q(z_2)$ if and only if $p(z_1) = p(z_2)$. Furthermore, $|q(z)| < 1$ for all $z \in D$.

(4) Let $g_2(z) = \dfrac{r^{\frac{1}{2}} - q(z)}{1 - r^{\frac{1}{2}}q(z)}$. The map g_2 is q followed by an automorphism of $\mathbb{D}$ that sends $r^{\frac{1}{2}}$ to 0.

(5) Let $h(z) = e^{i\theta} g_2(z)$. The map h is g_2 followed by a rotation through the angle θ.

Conclusion: h is a univalent mapping of D into $\mathbb{D}$.

We calculate $h(0)$ and $h'(0)$. In order to use the chain rule we need to see what happens to zero under all the maps used to construct h. It is easily checked that

$$g_1(0) = 0, \ p(0) = r, \ q(0) = r^{\frac{1}{2}}, \ g_2(0) = 0 \text{ and } h(0) = 0.$$

Aside: If $A(z) = \dfrac{\alpha z + \beta}{\gamma z + \delta}$, then $A'(z) = \dfrac{\alpha\delta - \beta\gamma}{(\gamma z + \delta)^2}$.

---

[2]By Exercise 5.1 of Chapter 5 there certainly exists a function q whose square is p. Hence $-q$ is also such a function. These are the two *branches* of the square root of p.

Now the calculation of the derivative of h at zero proceeds as follows.

$$h'(0) = e^{i\theta}g_2'(0) = e^{i\theta}\frac{r-1}{(1-r^{\frac{1}{2}}q(0))^2}q'(0) = e^{i\theta}\frac{r-1}{(1-r)^2}q'(0)$$

$$= \frac{e^{i\theta}}{r-1}\left(\frac{1}{2}\right)p(0)^{-\frac{1}{2}}p'(0) = \frac{e^{i\theta}}{r-1}\left(\frac{1}{2}\right)r^{-\frac{1}{2}}\frac{-1+r^2}{(1-rg_1(0))^2}g_1'(0)$$

$$= \frac{1}{2}e^{i\theta}\frac{1}{r-1}\frac{1}{r^{\frac{1}{2}}}\frac{r^2-1}{1}e^{-i\theta}\mu = \frac{r+1}{2r^{\frac{1}{2}}}\mu\ .$$

Finally,

$$\frac{r+1}{2r^{\frac{1}{2}}} > 1 \text{ if and only if } 0 < r < 1\,,$$

arriving at a contradiction that finishes the existence proof.

Uniqueness: The proof of uniqueness is a straightforward argument using the Schwarz's lemma. $\qquad\square$

COROLLARY 8.21. *If D is a simply connected domain in $\widehat{\mathbb{C}}$, then D is conformally equivalent to one and only one of the following: (i) $\widehat{\mathbb{C}}$, (ii) $\mathbb{C}$, or (iii) $\mathbb{D}$.*

The case (i), (ii), or (iii) occurs when the boundary of D consists of none, one, or more than one point, respectively. In the last case the boundary of D contains a continuum (a homeomorphic image of a closed interval containing more than one point).

8.4. Hyperbolic geometry

Let D be a simply connected domain in the extended complex plane with two or more boundary points. In this section we establish that such a domain carries a conformally invariant metric, known as the Poincaré or hyperbolic metric.

We consider only those simply connected domains D which, by the Riemann mapping theorem, are conformally equivalent to the upper half-plane $\mathbb{H}^2$, or, equivalently, to the unit disk $\mathbb{D}$, although the metric may be defined on all domains in $\widehat{\mathbb{C}}$ with two or more boundary points.

We show that conformal equivalences between these domains preserve the hyperbolic metric; that is, they are isometries (distance preserving maps) with respect to the hyperbolic metrics on the respective domains. Endowed with these equivalent metrics, the upper half-plane and the unit disk become models for non-Euclidean (also known as hyperbolic or Lobachevsky) geometry. As we have shown, the groups of conformal self-maps of these domains, $\mathrm{Aut}(\mathbb{H}^2)$ and $\mathrm{Aut}(\mathbb{D})$, consist of Möbius transformations, a class of maps much easier to study than the group of conformal self-maps of an arbitrary D. It is a remarkable fact that these Möbius functions constitute the full group of isometries of $\mathbb{H}^2$ and $\mathbb{D}$ with their respective hyperbolic metrics. We conclude this section using Schwarz's lemma and the hyperbolic metric to establish a deep connection between complex analysis and geometry. Namely, holomorphic maps between hyperbolic domains are either isometries or contractions with respect to their hyperbolic metrics.

We first define the Poincaré metric in a general setting, that is, on an arbitrary simply connected domain D with two or more boundary points (Section 8.4.1). We subsequently study it in more detail on $\mathbb{H}^2$ and $\mathbb{D}$, where specific computations are most easily carried out (Sections 8.4.2 and 8.4.3). The results that follow from these computations transfer to the general setting because of the conformal equivalence. Finally in Section 8.4.4, we establish the result about contractions.

8.4.1. The Poincaré metric.

DEFINITION 8.22. Let D be a simply connected domain in the extended complex plane with two or more boundary points. We define the *(infinitesimal form of the) Poincaré metric*

$$\lambda_D(z)\,|dz|$$

as follows. First, set for the unit disk

$$\lambda_{\mathbb{D}}(z) = \frac{2}{1 - |z|^2}, \ z \in \mathbb{D}.$$

For arbitrary D, choose a Riemann map $\pi : \mathbb{D} \to D$ and define λ_D by

$$\lambda_{\pi(\mathbb{D})}(\pi(z))\,|\pi'(z)| = \lambda_{\mathbb{D}}(z), \ z \in \mathbb{D}.$$

Our first task is to show that $\lambda_D(z)$ is well defined for all simply connected domains[3] D and all $z \in D$. Toward this end, let A be a conformal self-map of $\mathbb{D}$. Recall that there exist complex numbers a and b with $|a|^2 - |b|^2 = 1$ such that

$$A(z) = \frac{az + b}{\overline{b}z + \overline{a}}, \ z \in \mathbb{D}.$$

An easy calculation now shows that

$$\lambda_{\mathbb{D}}(A(z))\,|A'(z)| = \lambda_{\mathbb{D}}(z), \ z \in \mathbb{D}.$$

Let $z_0 \in D$ be arbitrary and suppose that π and ρ are two Riemann maps of $\mathbb{D}$ onto D with

$$\pi(z) = z_0 = \rho(\zeta)$$

---

[3]With two or more boundary points.

for two points z and $\zeta \in \mathbb{D}$. We need to show that[4]

$$\lambda_{\pi(\mathbb{D})}(\pi(z)) = \lambda_{\rho(\mathbb{D})}(\rho(\zeta)).$$

Now, $A = \pi^{-1} \circ \rho$ is in $\mathrm{Aut}(\mathbb{D})$ and $A(\zeta) = z$. It follows that

$$\lambda_{\rho(\mathbb{D})}(\rho(\zeta)) = \lambda_{\mathbb{D}}(\zeta) |\rho'(\zeta)|^{-1} = \lambda_{\mathbb{D}}(A(\zeta)) |A'(\zeta)| |\pi'(A(\zeta))|^{-1} |A'(\zeta)|^{-1}$$

$$= \lambda_{\mathbb{D}}(z) |\pi'(z)|^{-1} = \lambda_{\pi(\mathbb{D})}(\pi(z)).$$

REMARK 8.23. (1) If $z \in D$ is arbitrary and we choose the Riemann map π to satisfy $\pi(0) = z$, then

$$\lambda_D(z) = 2 |\pi'(0)|^{-1}.$$

(2) It is easy to see that

$$\lambda_{\mathbb{H}^2}(z) = \frac{1}{\Im z} \quad \text{for all } z \in \mathbb{H}^2.$$

The important invariance property of our metric is described by

PROPOSITION 8.24. *For every conformal map f defined on D,*

$$\lambda_{f(D)}(f(z)) |f'(z)| = \lambda_D(z) \quad \text{for all } z \in D.$$

PROOF. If π is a Riemann map, so is $f \circ \pi$. $\qquad\qquad\square$

Any infinitesimal metric on D allows us to define lengths of paths in D and, hence, a distance function on the domain. We work, of course, with length element

$$ds = \lambda_D(z) |dz|.$$

DEFINITION 8.25. We define *the hyperbolic length of a piecewise differentiable curve γ in D* by

$$l_D(\gamma) = \int_\gamma \lambda_D(z) |dz|;$$

if z_1 and z_2 are any two points in D, *the hyperbolic (or Poincaré) distance* between them by

$$\rho_D(z_1, z_2) = \inf\{l_D(\gamma); \gamma \text{ is a pdp in } D \text{ from } z_1 \text{ to } z_2\}. \qquad (8.3)$$

---

[4]The two symbols λ_D and $\lambda_{\pi(\mathbb{D})}$ denote, of course, the same function. The latter one is meant to emphasize the use of the Riemann map π in its computation.

We leave to the reader (Exercise 8.23) the verification that ρ_D defines a metric on D.

We remind the reader that an *isometry* from one metric space to another is a distance preserving map between them. It follows from the last proposition that for every conformal map f defined on D and every pdp γ in D,

$$l_{f(D)}(f \circ \gamma) = l_D(\gamma)$$

and

$$\rho_{f(D)}(f(z_1), f(z_2)) = \rho_D(z_1, z_2) \text{ for all } z_1 \text{ and } z_2 \in D;$$

that is, ρ is conformally invariant and an isometry with respect to the appropriate hyperbolic metrics.

8.4.2. Upper half-plane model. In this case $ds = \frac{|dz|}{\Im(z)}$. The hyperbolic length of an arbitrary curve γ in $\mathbb{H}^2$ may be hard to calculate directly from Formula (8.3). We will show that given any two points in $\mathbb{H}^2$, there exists a unique semicircle (a hyperbolic line) in $\mathbb{H}^2$ passing through the two points such that the hyperbolic length of its segment joining the two points realizes the hyperbolic distance between the two points. Such a hyperbolic line is called a *geodesic*, and the unique portion of the geodesic between the two points is called a *geodesic path* or *geodesic segment*. The following three lemmas establish the existence of a geodesic path between two points; the proof of its uniqueness follows.

LEMMA 8.26. *Let P and $Q \in \mathbb{H}^2$ lie on an Euclidean circle C centered on the real axis, and let γ be the arc of C in $\mathbb{H}^2$ between P and Q. Assume also that the radii from the center to P and Q make, respectively, angles α and β with the positive real axis.*
    Then

$$l_{\mathbb{H}^2}(\gamma) = \left| \log \frac{\csc(\beta) - \cot(\beta)}{\csc(\alpha) - \cot(\alpha)} \right|.$$

PROOF. Assume the circle C has radius r and is centered at $(c, 0)$ (see Figure 8.2).

Let $z = (x, y)$ be an arbitrary point on C, and let t be the angle that the radius from z to the center of C makes with the positive real axis; then $x = c + r \cos t$ and $y = r \sin t$. Therefore, $dx = -r \sin t$ and $dy = r \cos t$. We have $l_{\mathbb{H}^2}(\gamma) = \left| \int_\alpha^\beta \csc t \, dt \right|$, and the result follows. $\square$

Similarly one can calculate that

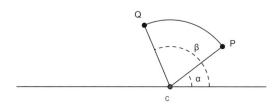

FIGURE 8.2. In the upper half-plane

LEMMA 8.27. *If $P = x_P + \imath y_P$ and $Q = x_P + \imath y_Q$, so that P and Q lie on a vertical line perpendicular to the $\mathbb{R}$-axis, and $y_P \geq y_Q > 0$, then the hyperbolic length of the segment of the vertical line connecting the points is given by $\log \frac{y_P}{y_Q}$.*

We recall that according to our conventions a straight line in $\mathbb{C}$ is a circle in $\mathbb{C} \cup \{\infty\}$ passing through infinity. By abuse of language, we will consider such circles to be centered on the real axis. Recall also that a Euclidean circle with center on the real axis is perpendicular to the real axis.

DEFINITION 8.28. For any circle C centered on the real axis, the part of C lying in the upper half-plane will be called a *semicircle* or a *hyperbolic line* or a *geodesic*.

LEMMA 8.29. *Let z and w be two different points in $\mathbb{H}^2$. Then there is a unique circle C centered on the real axis and passing through z and w.*

PROOF. If $\Re(z) = \Re(w)$, take C to be the Euclidean line through z and w. Otherwise, let L be the perpendicular bisector of the Euclidean line segment connecting z and w. If c is the point where L intersects the real line, then C is the circle with center c passing through z and w. See Figure 8.3. $\square$

COROLLARY 8.30. *Let z and w be two different points in $\mathbb{H}^2$. Then there is a unique geodesic passing through z and w.*

REMARK 8.31. If z and w are two different points in $\mathbb{H}^2$, we have observed in Lemma 8.29 that they determine a circle C centered on the real axis passing through z and w. We let z^* and w^* denote the points on $C \cap (\mathbb{R} \cup \{\infty\})$. We choose z^* so that it is closer to z than

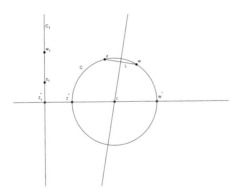

FIGURE 8.3. Unique circle through two points in the upper half-plane

to w. When $\Re(z) = \Re(w)$, z^* and w^* are $\Re(z)$ and ∞, respectively, if $\Im z < \Im w$; otherwise, $z^* = \infty$ and $w^* = \Re(w)$.

LEMMA 8.32. *Let z and w be two different points in $\mathbb{H}^2$. Then there exists a unique T in $\mathrm{Aut}(\mathbb{H}^2)$ such that $T(z^*) = 0$, $T(z) = \imath$, $T(w) = \imath y$ with $y > 1$, and $T(w^*) = \infty$.*

PROOF. Consider the unique circle C centered on the real axis and passing through z and w. Since the Möbius group is triply transitive, there exists a Möbius transformation T that maps z^*, z, w^* to 0, $\imath$, ∞, respectively. Since Möbius transformations map circles to circles (including straight lines), T maps C onto the imaginary axis and, since they preserve orthogonality, T maps $\mathbb{R} \cup \{\infty\}$ onto itself. Thus T is in $\mathrm{Aut}(\mathbb{H}^2)$. □

DEFINITION 8.33. Let z and w be two different points in $\mathbb{H}^2$. The arc of the unique geodesic determined by z and w between z and w is called the *geodesic segment* or *geodesic path* joining z and w.

LEMMA 8.34. *Let z and w be two different points in $\mathbb{H}^2$. Then the hyperbolic length of the geodesic segment joining z and w is shorter than the hyperbolic length of any other pdp γ joining z and w in $\mathbb{H}^2$.*

PROOF. Write $z = x_z + \imath y_z$ and $w = x_w + \imath z_w$.
First consider the case $x_z = x_w$. Assume the curve γ is given by $\gamma : [a, b] \to \mathbb{H}^2$, $\gamma(t) = x(t) + \imath y(t)$. Then

$$l_{\mathbb{H}^2}(\gamma) = \left| \int_a^b \frac{\sqrt{x'(t)^2 + y'(t)^2}}{y(t)}\, dt \right| \geq \left| \int_a^b \frac{\sqrt{y'(t)^2}}{y(t)} \right| = \left| \log \frac{y_w}{y_z} \right|,$$

and equality is attained if and only if $x(t)$ is constant and $y'(t) \geq 0$.

For the case $x_z \neq x_w$, by Lemma 8.32, we can find T in $\mathrm{Aut}(\mathbb{H}^2)$ such that $T(z^*) = 0$ and $T(w^*) = \infty$. Since T is conformal, the image under T of the geodesic segment between z and w is the segment on the imaginary axis between $T(z)$ and $T(w)$, and both segments have the same hyperbolic length. Similarly, $T \circ \gamma$ is a pdp in $\mathbb{H}^2$ joining $T(z)$ and $T(w)$, with the same hyperbolic length as γ, and we are reduced to the previous case. □

The next result follows from Lemmas 8.29 and 8.34.

THEOREM 8.35. *For any two points z and w in $\mathbb{H}^2$, the geodesic segment joining z to w is the unique curve that achieves the infimum for the hyperbolic metric defined by (8.3).*

Recall that (a, b, c, d) denotes the cross-ratio for any four points a, b, c, d in $\mathbb{C} \cup \{\infty\}$. A computation now establishes

LEMMA 8.36. *For any two points z and w in $\mathbb{H}^2$, the hyperbolic length of the geodesic segment γ joining z and w is given by*

$$\rho_{\mathbb{H}^2}(\gamma) = \log |(z^*, w^*, w, z)| = \log |(0, \infty, \imath y, \imath)| = \log y,$$

where y is the real number (> 1) given in Lemma 8.32.

PROOF. A fractional linear transformation preserves the cross-ratio of any four points. We have already seen that fractional linear transformations mapping $\mathbb{H}^2$ to itself are isometries. □

In particular, for any z and w in $\mathbb{H}^2$, we have

COROLLARY 8.37. $\rho_{\mathbb{H}^2}(z, w) = \log |(z^*, w^*, w, z)|.$

A calculation (see Exercise 8.19) shows that we also have

COROLLARY 8.38.

$$\rho_{\mathbb{H}^2}(z, w) = \log \frac{|z - \overline{w}| + |z - w|}{|z - \overline{w}| - |z - w|}. \tag{8.4}$$

We have already concluded (because conformal maps preserve the infinitesimal form of the hyperbolic metric) that $\mathrm{PSL}(2, \mathbb{R})$ acts as a group of isometries of $\mathbb{H}^2$. The current development for the hyperbolic metric together with the two facts that fractional linear transformations preserve the cross-ratio and map circles to circles, give an alternative proof that $\mathrm{Aut}(\mathbb{H}^2) = \mathrm{PSL}(2, \mathbb{R})$ consists of hyperbolic isometries of $\mathbb{H}^2$. Moreover, using this development we can prove the converse; that is, that every orientation preserving isometry of the upper half-plane with its hyperbolic metric is a Möbius transformation. We need a preliminary

PROPOSITION 8.39. *If f is an isometry of $\mathbb{H}^2$ that fixes the imaginary axis pointwise, then f is the identity map.*

PROOF. Let $z = x + \imath y$ and $f = u + \imath v$. For all positive real numbers t we have $\rho_{\mathbb{H}^2}(z, \imath t) = \rho_{\mathbb{H}^2}(f(z), f(\imath t)) = \rho_{\mathbb{H}^2}(u + \imath v, \imath t)$. We calculate using (8.4) that

$$\left[x^2 + (y - t)^2\right] v = \left[u^2 + (v - t)^2\right] y$$

for all positive t. Thus, $y = v$ and $x^2 = u^2$. Thus, $f(z) = z$ or $f(z) = -\overline{z}$. Since f is orientation preserving, we are done. $\square$

THEOREM 8.40. *The set of orientation preserving isometries of $\mathbb{H}^2$ with its hyperbolic metric is precisely the set of fractional linear transformations mapping $\mathbb{H}^2$ to itself; that is, $\mathrm{PSL}(2, \mathbb{R})$.*

PROOF. Suppose g is such an isometry. Then g maps hyperbolic lines to hyperbolic lines. Thus there is a fractional linear transformation f that preserves $\mathbb{H}^2$ and such that $f \circ g$ fixes the imaginary axis. Using (if needed) the isometries $z \mapsto kz$, k in $\mathbb{R}_{>0}$, and $z \mapsto \frac{-1}{z}$, we may assume that $f \circ g$ fixes $\imath$ and the intervals $(\imath, \infty)$ and $(0, \imath)$ on the imaginary axis.

Using the fact that $\rho_{\mathbb{H}^2}(\imath z, \imath w) = \left|\log\left(\frac{z}{w}\right)\right|$ for $z, w \in \mathbb{R}$ and $w \neq 0$, we see that $f \circ g = id$ on the imaginary axis, hence on $\mathbb{H}^2$, by the previous proposition. We conclude that g is a Möbius transformation. $\square$

8.4.3. Unit disk model. Statements about the hyperbolic metric on the upper half-plane can be translated to the unit disk model using the differential $ds = \dfrac{2\,|dz|}{1 - |z|^2}$. Most of these statements are left as exercises and do not need restatement. We emphasize the following two results.

In particular, Theorem 8.40 translates to

THEOREM 8.41. *The set of orientation preserving isometries of $\mathbb{D}$ with its hyperbolic metric is precisely the set of fractional linear transformations mapping $\mathbb{D}$ to itself, that is, $\mathrm{Aut}(\mathbb{D})$.*

And Corollary 8.38 becomes

THEOREM 8.42. *We have*

$$\rho_{\mathbb{D}}(w, z) = \log \frac{|1 - w\overline{z}| + |w - z|}{|1 - w\overline{z}| - |w - z|} \quad \textit{for all } z, w \in \mathbb{D}. \qquad (8.5)$$

In particular,

$$\rho_{\mathbb{D}}(0, z) = \log \frac{1 + |z|}{1 - |z|}, z \in \mathbb{D}. \tag{8.6}$$

PROOF. See Exercise 8.22.   □

8.4.4. Contractions and the Schwarz's lemma. We use the Schwarz's lemma to establish a deep connection between function theory and geometry. First we recall that not every holomorphic self-map of $\mathbb{D}$ is a Möbius transformation, only conformal self-maps are, and as we have seen, these are isometries in the hyperbolic metric. However,

THEOREM 8.43. *Holomorphic self-maps of the unit disk do not increase distance in the hyperbolic metric.*

PROOF. Let $F : \mathbb{D} \to \mathbb{D}$ be a nonconstant holomorphic function from the unit disk to itself. Assume first that $F(0) = 0$. Then by Schwarz's lemma (Theorem 5.34), we have $|F(z)| \leq |z|$ for $|z| < 1$ and $|F'(0)| \leq 1$.

Equation (8.6) also tells us that

$$\rho_{\mathbb{D}}(0, F(z)) = \log \frac{1 + |F(z)|}{1 - |F(z)|}.$$

Now we use Schwarz's lemma to conclude that

$$\frac{1 + |F(z)|}{1 - |F(z)|} \leq \frac{1 + |z|}{1 - |z|},$$

and then we conclude, by Equation (8.6), that

$$\rho_{\mathbb{D}}(0, F(z)) \leq \rho_{\mathbb{D}}(0, z).$$

Let a and $b \in \mathbb{D}$. Assume that $a \neq b$. Choose conformal self-maps A and B of the unit disk such that $B(0) = a$ and $A(F(a)) = 0$. Then $A \circ F \circ B$ is a holomorphic self-map of the unit disk that fixes 0. Hence, by what was already established,

$$\rho_{\mathbb{D}}(A(F(B(0))), A(F(B(z)))) \leq \rho_{\mathbb{D}}(0, z).$$

Since B and A are isometries,

$$\rho_{\mathbb{D}}(0, z) = \rho_{\mathbb{D}}(B(0), B(z))$$

and

$$\rho_{\mathbb{D}}(A(F(B(0))), A(F(B(z)))) = \rho_{\mathbb{D}}(F(B(0)), F(B(z))).$$

We conclude that

$$\rho_{\mathbb{D}}(F(B(0)), F(B(z))) \leq \rho_{\mathbb{D}}(B(0), B(z)).$$

Taking $z = B^{-1}(b)$, we obtain the required inequality

$$\rho_{\mathbb{D}}(F(a), F(b)) \leq \rho_{\mathbb{D}}(a, b). \tag{8.7}$$

If we multiply each side of the last equation by

$$\left| \frac{1}{a - b} \right| = \left| \frac{1}{F(a) - F(b)} \cdot \frac{F(a) - F(b)}{a - b} \right|$$

and take the limit as a approaches b, we get the infinitesimal form of our required formula

$$\lambda_{\mathbb{D}}(F(b)) \, |F'(b)| \leq \lambda_{\mathbb{D}}(b). \tag{8.8}$$

We leave it to the reader to verify that equality in either (8.7) or (8.8) implies that F is conformal (also an isometry). $\square$

DEFINITION 8.44. *Let M be a metric space with distance d. A map f from M to itself is a* contraction *if $d(f(x), f(y)) < d(x, y)$ for all x and y in M.*

Theorem 8.43 can be restated as

THEOREM 8.45. *A holomorphic self-map of the unit disk is either an isometry or a contraction with respect to the hyperbolic metric.*

8.5. Finite Blaschke products

Let $\mathbb{A} = \{a_0, a_1, \ldots\}$ be a finite or countable sequence of complex numbers lying in the unit disk $\mathbb{D}$. Define

$$B_{\mathbb{A}}(z) = B(z) = \prod_i \frac{|a_i|}{a_i} \cdot \frac{z - a_i}{1 - \overline{a_i} z}, \quad \text{for } z \in \mathbb{D},$$

where if $a_i = 0$, we set $\frac{|a_i|}{a_i} = 1$.

DEFINITION 8.46. *The function $B_{\mathbb{A}}$ is called a (finite or infinite)* Blaschke product associated with $\mathbb{A}$. *In the infinite case, there are, of course, convergence issues (see Section 10.5).*

For the rest of this section we will study *finite* Blaschke products. It follows immediately from the definition that we have

PROPOSITION 8.47. *Let $\mathbb{A} = \{a_0, a_1, \ldots, a_n\}$ be a finite sequence of points in $\mathbb{D}$. Then*
(a) $B = B_{\mathbb{A}}$ *is a meromorphic function on $\mathbb{C} \cup \{\infty\}$, with zeros precisely[5] at the $n + 1$ points $\{a_i\}$ and poles at the $n + 1$ points $\left\{ \frac{1}{\overline{a_i}} \right\}$, and*

---

[5]If a_i appears ν times in our list $\mathbb{A}$, then

$$\nu_{a_i}(B) = \nu \quad \text{and} \quad \nu_{\frac{1}{\overline{a_i}}}(B) = -\nu.$$

(b) *B is a self-map of the closed unit disk that maps the open unit disk holomorphically onto itself and the unit circle onto itself, with* $B(0) = (-1)^{n+1} \prod_i |a_i|$.

Blaschke products transform beautifully under automorphisms of $\mathbb{D}$, as shown next.

PROPOSITION 8.48. *Let* $\mathbb{A} = \{a_0, a_1, \cdots, a_n\}$ *be a finite sequence of points in* $\mathbb{D}$, *and let* T *be any element of* $\mathrm{Aut}(\mathbb{D})$. *Then*

$$B_{\mathbb{A}} \circ T = \lambda \, B_{T^{-1}(\mathbb{A})},$$

where λ *is a constant of absolute value 1 and*

$$T^{-1}(\mathbb{A}) = \{T^{-1}(a_0),\ T^{-1}(a_1),\ \cdots\}.$$

PROOF. Since T belongs to $\mathrm{Aut}(\mathbb{D})$, there exist complex numbers a and b, with $|a|^2 - |b|^2 = 1$, and such that $T(z) = \dfrac{a\,z + b}{\bar{b}\,z + \bar{a}}$ for all z in $\mathbb{D}$.

It suffices to compute the action of T on the function $z \mapsto \frac{z-\alpha}{1-\bar{\alpha}z}$, where $\alpha \in \mathbb{D}$. A calculation shows that

$$\frac{T(z) - \alpha}{1 - \bar{\alpha}T(z)} = \frac{z - \dfrac{\bar{a}\alpha - b}{-\bar{b}\alpha + a}}{\dfrac{\bar{a} - \bar{b}\bar{\alpha}}{a - \bar{b}\alpha}\left(1 - \dfrac{a\bar{\alpha} - \bar{b}}{-\bar{b}\bar{\alpha} + \bar{a}}z\right)}.$$

The proof is completed by observing that $\left|\dfrac{\bar{a} - \bar{b}\bar{\alpha}}{a - \bar{b}\alpha}\right| = 1$ and by recalling that $T^{-1}(w) = \dfrac{\bar{a}w - b}{-\bar{b}w + a}$. $\qquad\square$

THEOREM 8.49. *Let* f *be a holomorphic self-map of* $\mathbb{D}$. *Let* $\mathbb{A} = \{a_0, a_1, \cdots, a_n\}$ *be a finite sequence of points in* $\mathbb{D}$ *and* $B = B_{\mathbb{A}}$. *Assume that* $f(a_i) = 0$ *for each* a_i *in* $\mathbb{A}$. *Then*

(a) $|f(z)| \le |B(z)|$ *for all* $z \in \mathbb{D}$, *and*

(b) *if* $|f(a)| = |B(a)|$ *for some* $a \in \mathbb{D}$ *with* $a \ne a_i$ *for all* i, *then there is* $\lambda \in \mathbb{C}$ *with* $|\lambda| = 1$ *such that*

$$f(z) = \lambda B(z) \quad \text{for all } z \in \mathbb{D}. \tag{8.9}$$

(c) *If* a_i *appears* ν *times in the sequence* $\mathbb{A}$ *and if*

$$0 = f(a_i) = f'(a_i) = \ \ldots \ = f^{(\nu-1)}(a_i)$$

_____

.

and
$$\left|f^{(\nu)}(a_i)\right| = \left|B^{(\nu)}(a_i)\right|,$$
then there is $\lambda \in \mathbb{C}$ *with* $|\lambda| = 1$ *such that* (8.9) *holds.*

PROOF. Let T be an automorphism of $\mathbb{D}$ that sends 0 to a_0. Then, as a result of the last proposition, and replacing f and B by $f \circ T$ and $B \circ T$, respectively, we may assume that $a_0 = 0$.

We let B_0 be the Blaschke product associated with $\{a_1, a_2, \cdots, a_n\}$. The function $F = \dfrac{f}{B_0}$ is certainly holomorphic on $\mathbb{D}$ and $F(0) = 0$. We claim that $|F(z)| \leq 1$ for all $z \in \mathbb{D}$. Fix such a point z. The restrictions of $|B_0|$ to circles of radius r, with $0 \leq r \leq 1$, yields a family of functions that uniformly approach the constant function 1 as r approaches 1. Hence for all $\epsilon > 0$ we can choose an r such that $|B_0(w)| \geq 1 - \epsilon$ for all w of absolute value r. Without loss of generality, we can choose $r \geq |z|$. Hence, by the maximum principle for $|w| \leq r$, it follows that $|F(w)| = \dfrac{|f(w)|}{|B_0(w)|} \leq \dfrac{1}{1 - \epsilon}$. In particular, $|F(z)| \leq \frac{1}{1-\epsilon}$. Since ϵ is arbitrary, the claim is proved. By Schwarz's lemma, we obtain $|F(z)| \leq |z|$ or, equivalently,
$$|f(z)| \leq |z|\,|B_0(z)| = |B(z)|$$
for all $z \in \mathbb{D}$, proving (a).

The function $\dfrac{f}{B_{\mathbb{A}}}$ is analytic on $\mathbb{D}$, and its modulus is at most 1 on $\mathbb{D}$. Assuming $|f(a)| = |B(a)|$ for some $a \in \mathbb{D}$ with $a \neq a_i$ for all i, then there is a point a in $\mathbb{D}$ where the modulus of $\dfrac{f}{B_{\mathbb{A}}}$ is 1; therefore, it is constant, and (b) follows.

To prove (c), we may assume that $a_i = 0$. By L'Hopital's rule,
$$\left|\frac{f}{B_{\mathbb{A}}}(0)\right| = \left|\frac{f^{(\nu)}(0)}{B_{\mathbb{A}}^{(\nu)}(0)}\right| = 1,$$
and (8.9) follows.   □

Exercises

8.1. A matrix $A = \begin{bmatrix} a & b \\ c & d \end{bmatrix}$ with $a, b, c, d \in \mathbb{C}$, $ad - bc \neq 0$, acts on $\widehat{\mathbb{C}}$ by $z \mapsto \frac{az+b}{cz+d}$.
Show that for each $t \in \mathbb{C}_{\neq 0}$, A and tA induce the same action.

8.2. Suppose the four distinct points z_1, z_2, z_3, z_4 are permuted. What effect will this have on the cross-ratio (z_1, z_2, z_3, z_4)?

8.3. Prove in detail that the angles at z_1 and z_2 in Figure 8.1 are equal precisely when the four points lie on a circle in $\widehat{\mathbb{C}}$.

8.4. Show that $f \in \operatorname{Aut}(\mathbb{D})$ if and only if there exist θ in $\mathbb{R}$ and $\zeta \in \mathbb{D}$ such that

$$f(z) = e^{i\theta} \frac{z - \zeta}{1 - \bar{\zeta} z} \quad \text{for all } z \in \mathbb{D}.$$

8.5. Formulate and prove (as a consequence of the Riemann Mapping Theorem 8.20) the Riemann Mapping Theorem for simply connected domains $D \subseteq \widehat{\mathbb{C}}$. Include the possibility that $\zeta = \infty$.

8.6. (1) Let D be the domain in the extended complex plane $\widehat{\mathbb{C}}$ exterior to the circles $|z - 1| = 1$ and $|z + 1| = 1$. Find a Riemann Map (one-to-one holomorphic map) of D onto the strip $S = \{z \in \mathbb{C};\ 0 < \Im z < 2\}$.
(2) Find a conformal map from the domain in $\widehat{\mathbb{C}}$ given by

$$\{z \in \mathbb{C}; |z - 1| > 1, |z + 1| > 1\} \cup \{\infty\}$$

onto the upper half-plane.

8.7. Find a conformal map from $\{z \in \mathbb{C}; |z| < 1, \Im z > 0\}$ (a semi-disk) onto the unit disk.

8.8. If $f(z) = w$ is a Riemann Map from the domain $|\operatorname{Arg} z| < \frac{\pi}{100}$ onto the domain $|w| < 1$ and if $f(1) = 0$ and $f'(1) > 0$, find $f(2)$.

8.9. Find a conformal map from the disk $\{z \in \mathbb{C}; |z| < 1\}$ onto $\{z \in \mathbb{C}; |z| < 1, \Im z > 0\}$.

8.10. If $w = g(z)$ maps the quadrant $\{z = x + iy \in \mathbb{C}; x > 0, y > 0\}$ conformally onto $|w| < 1$ with $g(1) = 1$, $g(i) = -1$, and $g(0) = -i$, find $|g'(1 + i)|$.

8.11. If f is holomorphic for $|z| < 1$ and satisfies $|f(z)| < 1$ for $|z| < 1$ and $f(0) = f\left(\frac{1}{2}\right) = 0$, show that

$$|f(z)| \le \left| z \cdot \frac{2z - 1}{2 - z} \right| \quad \text{for all } |z| < 1.$$

8.12. For each $n = 1, 2, 3, \ldots$, find a conformal map from the infinite angular sector $0 < \operatorname{Arg} z < \frac{\pi}{n}$ onto the unit disk.

8.13. Find the Riemann Map f from the strip $0 < \Im z < 1$ onto the unit disk satisfying $f\left(\frac{i}{2}\right) = 0$ and $f'\left(\frac{i}{2}\right) > 0$.

8.14. Find a conformal map from the domain

$$\{z \in \mathbb{C}; |z| < 1 \text{ and } z \neq t \text{ for } 0 \leq t < 1\}$$

onto $|w| < 1$.

8.15. Find a conformal map from the upper half-plane onto the unit disk minus the nonnegative real numbers.

8.16. Suppose $\{f_n\}$ is a sequence of holomorphic functions in $|z| < 1$ that satisfy

$$\Re f_n(z) > 0 \quad \text{and} \quad |f_n(0) - \imath| < \frac{1}{2} \quad \text{for all } n \in \mathbb{Z}_{>0} \text{ and all } |z| < 1.$$

Show that $\{f_n\}$ contains a subsequence that converges uniformly on compact subsets of the unit disk.

8.17. Supply all of the details of the proof that $\rho_{\mathbb{H}^2}$ defines a metric that is invariant under the Möbius group $\text{PSL}(2, \mathbb{R})$.

8.18. Let z and w be points in $\mathbb{H}^2$ with z^* and w^* the points on the real axis that are the ends of the diameter of the circle perpendicular to the real lines passing through z and w. Define $D(z, w) = |\log(z^*, w^*, w, z)|$. Show that D defines a metric in the upper half-plane.

8.19. Let z and w be points in the upper half-plane. Prove that

$$\rho_{\mathbb{H}^2}(z, w) = \log \frac{|z - \overline{w}| + |z - w|}{|z - \overline{w}| - |z - w|}.$$

8.20. Show that the hyperbolic circle $\{z \in \mathbb{H}^2; \rho_{\mathbb{H}^2}(\imath, z) = r\}$ is given by

$$\{z \in \mathbb{H}^2; x^2 + (y - \cosh r)^2 = \sinh^2(r)\},$$

and conclude that the topology induced by the hyperbolic metric in $\mathbb{H}^2$ coincides with the Euclidean topology.

8.21. Assume f is a bounded holomorphic function on the unit disk and $f\left(\frac{\imath}{2}\right) = f\left(-\frac{\imath}{2}\right) = 0$. Show that

$$f(z) = \frac{z - \frac{\imath}{2}}{1 + \frac{\imath}{2}z} \frac{z + \frac{\imath}{2}}{1 - \frac{\imath}{2}z} G(z), \quad z \in \mathbb{D},$$

where G is a bounded holomorphic function on $\mathbb{D}$.

8.22. Prove that

$$\rho_{\mathbb{D}}(0, z) = \log \frac{1 + |z|}{1 - |z|}, \quad \text{for all } z \in \mathbb{D}.$$

8.23. Let D be a simply connected domain in the extended complex plane with two or more boundary points. Show that (8.3) defines a metric on D.

CHAPTER 9

Harmonic Functions

This chapter is devoted to the study of harmonic functions. These functions are closely connected to holomorphic maps since the real and imaginary parts of a holomorphic function are harmonic functions. The study of harmonic functions is important in physics and engineering, and there are many results in the theory of harmonic functions that are not connected directly with complex analysis. However, in this chapter we consider that part of the theory of harmonic functions that grows out of the Cauchy Theory. Mathematically this is quite pleasing. One of the most important aspects of harmonic functions is that they arise as functions that solve a *boundary value problem* for holomorphic functions, known as the Dirichlet problem. An example is the problem of finding a function continuous in a closed disk that assumes certain known values on the boundary of the disk and is harmonic in the interior of the disk. An important tool in the solution is the Poisson formula.

In the first section we define harmonic functions and the Laplacian of a function. In the second we obtain integral representations for harmonic functions that are analogous to the Cauchy Integral Formula, including the Poisson formula; in the third we use these integral representations to solve the Dirichlet problem. The third section includes three interpretations of the Poisson formula: a geometric interpretation, a Fourier series interpretation, and a classic one. In the fourth section we characterize harmonic functions by their Mean Value Property. The last section deals with the reflection principle for holomorphic and real-valued harmonic functions, which is a simple but useful extension tool.

9.1. Harmonic functions and the Laplacian

We begin with

DEFINITION 9.1. Let D be a domain in $\mathbb{C}$ and $g \in \mathbf{C}^2(D)$. We define $\triangle g$, the *Laplacian of g*, by

$$\triangle g = \frac{\partial^2 g}{\partial x^2} + \frac{\partial^2 g}{\partial y^2}. \tag{9.1}$$

The operator $\triangle$ is called the *Laplacian* or the *Laplace operator*.

DEFINITION 9.2. Let D be a domain in $\mathbb{C}$ and $g \in \mathbf{C}^2(D)$. We say that g is *harmonic* if

$$\triangle g = \frac{\partial^2 g}{\partial x^2} + \frac{\partial^2 g}{\partial y^2} = 0 \text{ in } D. \qquad \text{(Laplace)}$$

These definitions have several immediate consequences that we list:

(1) It is obvious from the definition of the Laplacian as a linear operator on C^2 complex-valued functions that it preserves real-valued functions. It is useful to have equivalent formulas for it (Exercise 9.1):

$$\triangle = \frac{\partial^2}{\partial x^2} + \frac{\partial^2}{\partial y^2} = 4\frac{\partial^2}{\partial \bar{z} \partial z} = 4\frac{\partial^2}{\partial z \partial \bar{z}} \qquad (9.2)$$

and in polar coordinates (r, θ)

$$\triangle = \frac{1}{r^2}\left(r\frac{\partial}{\partial r}\left(r\frac{\partial}{\partial r} \right) + \frac{\partial^2}{\partial \theta^2} \right). \qquad (9.3)$$

(2) Recall that for $f \in \mathbf{C}^1(D)$, f is holomorphic on D if and only if $\dfrac{\partial f}{\partial \bar{z}} = 0$ in D. Thus, for $f \in \mathbf{C}^2(D)$, f is harmonic if and only if $\dfrac{\partial f}{\partial z}$ is holomorphic.

   In particular, holomorphic functions are harmonic, and (2) gives an easy way to construct analytic functions from harmonic ones.

(3) f is harmonic if and only if $\bar{f}$ is.

(4) f is harmonic if and only if $\Re f$ and $\Im f$ are [this follows from the linearity of $\triangle$ and (3)].

(5) If f is holomorphic on D, then f, $\Re f$, $\Im f$ and $\bar{f}$ are harmonic on D.

(6) If f is holomorphic or anti-holomorphic on D and g is harmonic on $f(D)$, then $g \circ f$ is harmonic on D.

   PROOF. Assume that f is holomorphic, let $w = f(z)$, and use the chain rule (see Exercise 2.10):

$$(g \circ f)_z = g_w f_z + g_{\bar{w}} \bar{f}_z = g_w f_z$$

and

$$(g \circ f)_{z\bar{z}} = g_{ww} f_{\bar{z}} f_z + g_{w\bar{w}} \overline{f_{\bar{z}}} f_z + g_w f_{z\bar{z}} = 0.$$

The argument in the anti-holomorphic case is similar. □

(7) If $f \in \mathbf{C}^2(D)$ and f is locally the real part of an analytic function on D, then f is harmonic on D.

EXAMPLE 9.3. $\log|z|$ is harmonic on $\mathbb{C}_{\neq 0}$, since it is locally the real part of $\log z$, a multivalued, but holomorphic, function.

PROPOSITION 9.4. *If g is real valued and harmonic, then it is locally the real part of an analytic function. The analytic function is unique up to an additive constant.*

PROOF. Let D be a simply connected region where g is harmonic. Since $2g_z\, dz$ is closed on D, it is exact. Choose a holomorphic function f on D with $df = 2g_z\, dz$. Then

$$d\bar{f} = 2g_{\bar{z}}\, d\bar{z}$$

and hence,

$$\frac{1}{2} d(f + \bar{f}) = dg\,;$$

that is,

$$\Re f = g + \text{ constant}.$$

□

COROLLARY 9.5. *A real-valued harmonic function on a simply connected domain is the real part of a holomorphic function.*

COROLLARY 9.6. *A harmonic function is C^∞.*

COROLLARY 9.7. *Harmonic functions have the Mean Value Property, and hence they satisfy the Maximum Modulus Principle. Real-valued harmonic functions also satisfy the Maximum and Minimum Principles.*

PROOF. See Definition 5.28 and the properties that follow thereof.

□

REMARK 9.8. The Maximum (Minimum) Principle asserts that if f is real valued and harmonic on a domain D and if f has a relative maximum (minimum) at a point $\zeta \in D$, then f is constant in a neighborhood of ζ. Furthermore, if D is bounded and f is continuous on the closure of D, with $m \leq f \leq M$ on ∂D for some real constants m and M, then $m \leq f \leq M$ on D.

9.2. Integral representation of harmonic functions

We begin to apply the Cauchy Theory toward our present main goal, which is to solve a boundary value problem. A major tool in the solution is the *Poisson formula*. Given a harmonic function g defined in a disk, we derive an integral formula for g, known as the *Poisson formula*.

PROPOSITION 9.9 (**The Poisson formula**). *Let g be a harmonic function on the domain $|z| < \rho$ for some $\rho > 0$. Then, for each $0 < r < \rho$,*

$$g(z) = \frac{1}{2\pi} \int_0^{2\pi} g(re^{i\theta}) \cdot \frac{r^2 - |z|^2}{|re^{i\theta} - z|^2} \, d\theta \quad \text{for } |z| < r. \quad (9.4)$$

PROOF. It suffices to assume that g is real valued. To establish this formula, we can thus apply Proposition 9.4 and choose the holomorphic function f on this domain with $\Re f = g$ and $g(0) = f(0)$, noting that there is a unique such f.

The function f has a power series expansion: Let $0 < r < \rho$ and $z = re^{i\theta}$. Then

$$f(r\,e^{i\theta}) = f(z) = \sum_{n=0}^{\infty} a_n z^n, \quad \text{with } a_0 \in \mathbb{R}.$$

Now

$$g(z) = \frac{1}{2} \left(f(z) + \overline{f(z)} \right) = \frac{1}{2} \sum_{n=0}^{\infty} \left(a_n z^n + \overline{a_n z^n} \right)$$

$$= a_0 + \frac{1}{2} \sum_{n=1}^{\infty} r^n \left(a_n e^{i n\theta} + \overline{a_n}\, e^{-i n\theta} \right).$$

Integration of g along the curve $\gamma(\theta) = r\,e^{i\theta}$, for $0 \leq \theta \leq 2\pi$, yields

$$a_0 = \frac{1}{2\pi} \int_0^{2\pi} g(r\,e^{i\theta}) \, d\theta.$$

Multiplying g by $e^{-in\theta}$ for $n \in \mathbb{Z}_{>0}$ and integrating along the same curve, we obtain

$$\frac{1}{2} r^n a_n = \frac{1}{2\pi} \int_0^{2\pi} g(re^{i\theta}) \cdot e^{-in\theta} \, d\theta;$$

or, equivalently,

$$a_n = \frac{1}{\pi} \int_0^{2\pi} \frac{g(re^{i\theta})}{(re^{i\theta})^n} \, d\theta \,, \quad \text{for } n \geq 1.$$

Thus for $|w| < r$, we have

$$f(w) = \frac{1}{2\pi} \int_0^{2\pi} g(re^{i\theta}) \cdot \left[1 + 2 \sum_{n \geq 1} \left(\frac{w}{re^{i\theta}} \right)^n \right] d\theta .$$

Now

$$1 + 2 \sum_{n \geq 1} \left(\frac{w}{re^{i\theta}} \right)^n = \frac{re^{i\theta} + w}{re^{i\theta} - w}$$

and thus

$$f(w) = \frac{1}{2\pi} \int_0^{2\pi} g(re^{i\theta}) \cdot \frac{re^{i\theta} + w}{re^{i\theta} - w} \, d\theta \quad \text{for } |w| < r.$$

The above formula gives a representation of a holomorphic function in terms of its real part, when the function is real at 0. Taking the real part of both sides and renaming the variable w to z, we obtain equation (9.4), the Poisson formula. □

The function

$$\frac{r^2 - |z|^2}{|re^{i\theta} - z|^2} = \Re \left(\frac{re^{i\theta} + z}{re^{i\theta} - z} \right) \qquad \text{(Poisson kernel)}$$

is known as the *Poisson kernel*. Note that setting $z = 0$ in Formula (9.4), we reobtain the *Mean Value Property for harmonic functions*.

The derivation of the above formula assumed that g was harmonic in the closed disk $\{|z| \leq r\}$. However, the result remains true for $|z| < 1$ under the weaker assumption that g is harmonic in the open disk $\{|\zeta| < r\}$ and continuous on its closure. In this case, fix t with $0 < t < 1$ and look at the function of z given by $g(tz)$. It is harmonic on the closed disk $\{|\zeta| \leq r\}$, and hence, by the result already proven (9.4),

$$g(tz) = \frac{1}{2\pi} \int_0^{2\pi} \frac{r^2 - |z|^2}{|re^{i\theta} - z|^2} \cdot g(tre^{i\theta}) \, d\theta .$$

Since the function g is uniformly continuous on the closed disk, we know that $g(tz)$ approaches $g(z)$ uniformly on the circle $\{|\zeta| = r\}$ as t approaches 1. Hence both sides of the last equation converge to the expected quantities.

As a special case we apply the Poisson formula to the function that is identical to 1 and obtain

$$\int_0^{2\pi} \frac{r^2 - |z|^2}{|re^{i\theta} - z|^2} \, d\theta = 2\pi \quad \text{for all } z \in \mathbb{C} \text{ with } |z| < r .$$

DEFINITION 9.10. A *harmonic conjugate* of a real-valued harmonic function u is any real-valued function v such that $u + \imath v$ is holomorphic.

Harmonic conjugates always exist locally, and they exist globally on simply connected domains. They are unique up to additive real constants. In fact, it is easy to see that they are given locally as follows.

PROPOSITION 9.11. *If g is harmonic and real valued in $|z| < \rho$ for some $\rho > 0$, then the harmonic conjugate of g vanishing at the origin is given by*

$$\frac{1}{2\pi\imath} \int_0^{2\pi} g(re^{\imath\theta}) \cdot \frac{re^{-\imath\theta}z - re^{\imath\theta}\overline{z}}{|re^{\imath\theta} - z|^2} \, d\theta \,, \quad \text{for } |z| < r < \rho.$$

The following result is interesting and useful.

THEOREM 9.12 (**Harnack's inequalities**). *If g is a positive harmonic function on $|z| < r$ that is continuous on $|z| \leq r$, then*

$$\frac{r - |z|}{r + |z|} \cdot g(0) \leq g(z) \leq \frac{r + |z|}{r - |z|} \cdot g(0) \,, \quad \text{for all } |z| < r.$$

PROOF. Our starting point is (9.4). We use elementary estimates for the Poisson kernel:

$$\frac{r - |z|}{r + |z|} = \frac{r^2 - |z|^2}{(r + |z|)^2} \leq \frac{r^2 - |z|^2}{|re^{\imath\theta} - z|^2} \leq \frac{r^2 - |z|^2}{(r - |z|)^2} = \frac{r + |z|}{r - |z|}.$$

Multiplying these inequalities by the positive number $g(w) = g(r\,e^{\imath\theta})$ and then averaging the resulting function over the circle $|w| = r$, we obtain

$$\frac{r - |z|}{r + |z|} \cdot \frac{1}{2\pi} \int_0^{2\pi} g(re^{\imath\theta}) \, d\theta \leq \frac{1}{2\pi} \int_0^{2\pi} g(re^{\imath\theta}) \cdot \frac{r^2 - |z|^2}{|re^{\imath\theta} - z|^2} \, d\theta$$

$$\leq \frac{r + |z|}{r - |z|} \cdot \frac{1}{2\pi} \int_0^{2\pi} g(re^{\imath\theta}) \, d\theta.$$

The middle term in the above inequalities is $g(z)$ as a consequence of (9.4), whereas the extreme averages are $g(0)$ by the Mean Value Property. $\qquad\square$

REMARK 9.13. Exercise 9.6 gives a remarkable consequence of Harnak's inequalities.

9.3. The Dirichlet problem

Let D be a bounded region in $\mathbb{C}$, and let $f \in \mathbf{C}(\partial D)$. The *Dirichlet problem* is to find a continuous function u on the closure of D whose restriction to D is harmonic and which agrees with f on the boundary of D.

We will consider only the special case where D is a disk; without loss of generality, the disk has radius 1 and center 0.

For a piecewise continuous function u on S^1 and $z \in \mathbb{C}$ with $|z| < 1$, we define [compare with (9.4)]

$$P[u(z)] = \frac{1}{2\pi} \int_0^{2\pi} \Re\left(\frac{e^{i\theta} + z}{e^{i\theta} - z}\right) \cdot u(e^{i\theta})\, d\theta. \tag{9.5}$$

We have

$$P[u(z)] = \frac{1}{2\pi} \int_0^{2\pi} u(e^{i\theta}) \cdot \frac{1 - |z|^2}{|e^{i\theta} - z|^2}\, d\theta. \tag{9.6}$$

The following are properties of P:

(1) $P[u]$ is a well-defined function on the open unit disk. Hence we may view P as an operator that assigns a function, which we also denote by Pu, on the open unit disk to each piecewise continuous function u on the unit circle.

(2) $P[u + v] = Pu + Pv$ and $P[cu] = c \cdot Pu$ for all piecewise continuous functions u and v on S^1 and every constant c (P is a linear operator).

(3) $P[1] = 1$ because the constant function is analytic.

(4) $P[u]$ is harmonic in the open unit disk. To establish this claim we may assume (by linearity of the operator P) that u is real valued. In this case, $P[u]$ is obviously the real part of an analytic function on the disk.

(5) For all constants c,

$$P[c] = c.$$

This last fact (5) implies that any bound on u yields the same bound on Pu. For example, for real-valued u with $m \leq u \leq M$ for real constants m and M, we have $m \leq Pu \leq M$.

We now establish the solvability of the Dirichlet problem for disks.

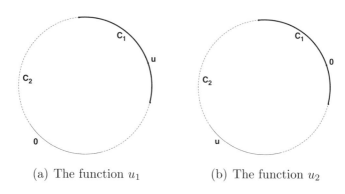

(a) The function u_1 (b) The function u_2

FIGURE 9.1. u_1 and u_2

THEOREM 9.14 (**H. A. Schwarz**). *Assume that u is a piecewise continuous function on the unit circle S^1.*

Then the function Pu is harmonic on $\{|z| < 1\}$ and, for $\theta_0 \in \mathbb{R}$, its limit as z approaches $e^{i\theta_0}$ is $u(e^{i\theta_0})$ provided u is continuous at $e^{i\theta_0}$. In particular, the Dirichlet problem is solvable for disks.

PROOF. We only have to study the boundary values for Pu.

Let C_1 and C_2 be complementary arcs on the unit circle. Let u_1 be the function which coincides with u on C_1 and vanishes on C_2; let u_2 be the corresponding function for C_2 (see Figure 9.1). Clearly $Pu = Pu_1 + Pu_2$.

The function Pu_1 can be regarded as an integral over the arc C_1; hence it is harmonic on $\mathbb{C} - C_1$. The expression

$$\Re\left(\frac{e^{i\theta} + z}{e^{i\theta} - z}\right) = \frac{1 - |z|^2}{|e^{i\theta} - z|^2}$$

vanishes on $|z| = 1$ for $z \neq e^{i\theta}$. It follows that Pu_1 is zero on the one-dimensional interior of the arc C_2. By continuity $Pu_1(z)$ approaches zero as z approaches a point in the interior of C_2.

In proving that Pu has limit $u(e^{i\theta_0})$ at $e^{i\theta_0}$, we may assume that $u(e^{i\theta_0}) = 0$ (if not replace u by $u - u(e^{i\theta_0})$). Under this assumption, given an $\epsilon > 0$ we can find C_1 and C_2 such that $e^{i\theta_0}$ is an interior point of C_2 and $\left|u(e^{i\theta})\right| < \frac{\epsilon}{2}$ for $e^{i\theta} \in C_2$. This last condition implies that $\left|u_2(e^{i\theta})\right| < \frac{\epsilon}{2}$ for all $e^{i\theta}$, and hence $|Pu_2(z)| < \frac{\epsilon}{2}$ for all $|z| < 1$.

But we also have that u_1 is continuous and vanishes at $e^{i\theta_0}$. Since Pu_1 is continuous at $e^{i\theta_0}$ and agrees with u_1 there, there exists a $\delta > 0$ such that $|Pu_1(z)| < \frac{\epsilon}{2}$ for $\left|z - e^{i\theta_0}\right| < \delta$.

It follows that $|Pu(z)| \leq |Pu_1(z)| + |Pu_2(z)| < \epsilon$ as long as $|z| < 1$ and $\left|z - e^{i\theta_0}\right| < \delta$. This is the required continuity statement.   □

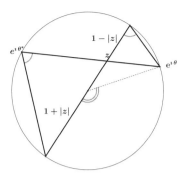

FIGURE 9.2. The similar triangles

9.3.1. Geometric interpretation of the Poisson formula.

This interpretation is due to Schwarz; the presentation follows Ahlfors.[1]
Recipe: *To find $P[u(z)]$ we replace $u(e^{i\theta})$—the value of the function u at the point $e^{i\theta}$—by its value $u(e^{i\theta^*})$ at the point $e^{i\theta^*}$ on the unit circle opposite to the point to $e^{i\theta}$ with respect to z, and we average these values over the unit circle (see Figure 9.2).*

This comes from reinterpreting the second formula for $P[u(z)]$, equation (9.6), as follows: Fix a point z inside the unit circle and a point $e^{i\theta}$ on the unit circle. Let $e^{i\theta^*}$ be the unique point on the unit circle that also lies on the straight line through z and $e^{i\theta}$. High-school geometry (similar triangles, see Figure 9.2) or a calculation (using the law of cosines, for example) yields

$$1 - |z|^2 = \left| e^{i\theta} - z \right| \cdot \left| e^{i\theta^*} - z \right|.$$

The ratio $\dfrac{e^{i\theta} - z}{e^{i\theta^*} - z}$ is negative; it follows from this observation that

$$1 - |z|^2 = -(e^{i\theta} - z) \cdot (e^{-i\theta^*} - \overline{z}). \qquad (9.7)$$

To verify this equality, note that

$$\frac{e^{i\theta} - z}{e^{i\theta^*} - z} \cdot (e^{i\theta^*} - z) \cdot (e^{-i\theta^*} - \overline{z})$$

is negative and has the same absolute value as $(e^{i\theta} - z) \cdot (e^{i\theta^*} - z)$.

We now regard θ^* as a function of θ, with z fixed, and differentiate equality (9.7) logarithmically to obtain

$$\frac{e^{i\theta}}{e^{i\theta} - z} \, d\theta = \frac{e^{-i\theta^*}}{e^{-i\theta^*} - \overline{z}} \, d\theta^*.$$

---

[1] *Complex Analysis* (third edition), McGraw-Hill, 1979.

Hence we see that (because θ^* is an increasing function of θ) that

$$\frac{d\theta^*}{d\theta} = \left| \frac{e^{i\theta^*} - z}{e^{i\theta} - z} \right| = \frac{1 - |z|^2}{|e^{i\theta} - z|^2} \,.$$

We have thus shown that

$$P[u(z)] = \frac{1}{2\pi} \int_0^{2\pi} u(e^{i\theta}) \, d\theta^* = \frac{1}{2\pi} \int_0^{2\pi} u(e^{i\theta^*}) \, d\theta.$$

The recipe follows.

9.3.2. Fourier series interpretation of the Poisson formula.
We again consider the case of the unit disk and proceed to compute the power series expansion of $Pu(z)$ at the origin.

We note that

$$P[u(z)] = \frac{1}{2\pi} \int_0^{2\pi} \frac{1 - z\overline{z}}{(e^{i\theta} - z) \cdot (e^{-i\theta} - \overline{z})} \cdot u(e^{i\theta}) \, d\theta \ \ \text{for } |z| < 1.$$

Start with an expansion of the Poisson kernel

$$\frac{1 - z\overline{z}}{(e^{i\theta} - z) \cdot (e^{-i\theta} - \overline{z})} = \frac{1 - z\overline{z}}{(1 - e^{-i\theta}z) \cdot (1 - e^{i\theta}\overline{z})}$$

$$= (1 - z\overline{z}) \sum_{n,m\geq 0} e^{-in\theta} \cdot z^n \cdot e^{im\theta}\overline{z}^m$$

$$= (1 - z\overline{z}) \sum_{n,m\geq 0} e^{i(m-n)\theta} z^n \overline{z}^m$$

$$= 1 + \sum_{n=1}^{\infty} e^{-in\theta} z^n + \sum_{m=1}^{\infty} e^{im\theta}\overline{z}^m \,,$$

and note that the last two series converge uniformly and absolutely on all compact subsets of the unit disk. Therefore we see that

$$P[u(z)] = a_0 + \sum_{n=1}^{\infty} a_n z^n + \sum_{m=1}^{\infty} b_m \overline{z}^m,$$

where, for $n \in \mathbb{Z}_{\geq 0}$ and $m \in \mathbb{Z}_{>0}$,

$$a_n = \frac{1}{2\pi} \int_0^{2\pi} e^{-in\theta} \cdot u(e^{i\theta}) \, d\theta \ , \ b_m = \frac{1}{2\pi} \int_0^{2\pi} e^{im\theta} \cdot u(e^{i\theta}) \, d\theta. \quad (9.8)$$

Thus we have the following procedure for extending a given continuous function u on the unit circle to a continuous function on the closed unit disk that is harmonic on its interior. First compute the Fourier series of u:

$$u(e^{i\theta}) = \sum_{n=0}^{\infty} a_n e^{in\theta} + \sum_{m=1}^{\infty} b_m e^{-im\theta},$$

where the Fourier coefficients a_n and b_m are given by (9.8). In this series, replace $e^{in\theta}$ by z^n, for each $n \in \mathbb{Z}_{\geq 0}$, and $e^{-im\theta}$ by $\bar{z}^m$, for each $m \in \mathbb{Z}_{>0}$.

9.3.3. Classic reformulation of the Poisson formula.
To give the classic reformulation, we need to define the *conjugate differential* of a given differential.

DEFINITION 9.15. If $\omega = P\,dx + Q\,dy$ is a differential form, we define $^*\omega$, the *conjugate differential of* ω, by

$$^*\omega = -Q\,dx + P\,dy.$$

If D is a simply connected domain in $\mathbb{C}$ and $u \in \mathbf{C}^2(D)$ is real valued, we know that u is harmonic on D if and only if u is the real part of an analytic function f on D. In this case,

$$df = f'(z)\,dz = (u_x - iu_y) \cdot (dx + i\,dy)$$
$$= (u_x\,dx + u_y\,dy) + i\,(-u_y\,dx + u_x\,dy) = du + i\,^*du$$

is an exact differential on D and $^*du = dv$, where v is a harmonic conjugate of u on D. Thus du and $^*du$ are exact differential forms on D whenever u is a real-valued harmonic function on a simply connected domain D.

In what follows we work with cycles rather than with curves. We remind the reader of the definitions of cycles and cycles homologous to zero in Section 5.2. In the general case, for a harmonic (including complex-valued) function u on an arbitrary (not necessarily simply connected) domain D, the form $du = u_x dx + u_y dy$ is always exact on D and its conjugate differential $^*du = -u_y dx + u_x dy$ is closed since $(-u)_{yy} = u_{xx}$. We conclude that

$$\int_{\gamma} {}^*du = 0 \tag{9.9}$$

for all harmonic functions u on D and all cycles γ in D that are homologous to zero on that domain.

We can now turn to the reformulation.

Assume that γ is a *regular* curve with equation $z = z(t)$ (regular means that $z'(t) \neq 0$ for all t). The direction of the tangent line to the curve at $z(t)$ is determined by the angle $\alpha = \operatorname{Arg} z'(t)$ and

$$dx = |dz| \cos \alpha \; , \; dy = |dz| \sin \alpha.$$

The *normal* that points to the right of the tangent line has direction $\beta = \alpha - \frac{\pi}{2}$. The *normal derivative* of u is the directional derivative of u in the direction β:

$$\frac{\partial u}{\partial n} = u_x \cos \beta + u_y \sin \beta = u_x \sin \alpha - u_y \cos \alpha.$$

Thus we see that $^*du = \frac{\partial u}{\partial n} |dz|$ and (9.9) can be rewritten as

$$\int_\gamma \frac{\partial u}{\partial n} |dz| = 0.$$

It is important to realize that if γ is the circle $|z| = r$, then $\dfrac{\partial u}{\partial n} = \dfrac{\partial u}{\partial r}$.

We now prove an important generalization of (9.9).

THEOREM 9.16. *If u_1 and u_2 are harmonic functions on D, then*

$$u_1 {}^*du_2 - u_2 {}^*du_1$$

is a closed form on D.

To establish this assertion, it involves no loss of generality to assume that the functions are real valued (see Exercise 9.4), and hence, we may also assume (because the issue is local) that each function u_j has a single-valued harmonic conjugate v_j; thus

$$u_1 {}^*du_2 - u_2 {}^*du_1 = u_1 \, dv_2 - u_2 \, dv_1 = u_1 \, dv_2 + v_1 \, du_2 - d(u_2 \, v_1).$$

The last expression $d(u_2 v_1)$ is, of course, exact and

$$u_1 \, dv_2 + v_1 \, du_2 = \Im \left((u_1 + \imath v_1)(du_2 + \imath \, dv_2) \right).$$

Now $u_1 + \imath v_1$ is an analytic function and $du_2 + \imath \, dv_2$ is the total differential of an analytic function. By Cauchy's theorem, the product is closed, and hence, we have shown that

$$\int_\gamma u_1 {}^*du_2 - u_2 {}^*du_1 = 0$$

for all cycles γ that are homologous to zero in D.

In classic language, the above formula reads as follows

$$\int_\gamma \left(u_1 \frac{\partial u_2}{\partial n} - u_2 \frac{\partial u_1}{\partial n} \right) |dz| = 0.$$

Let us take for D the annulus $\{z \in \mathbb{C}; R_1 < |z| < R_2\}$ and apply the above formula to the function $z \mapsto u_1(z) = \log r$ (in polar coordinates) and an arbitrary harmonic function $u_2 = u$ on D. We take for γ the cycle $C_1 - C_2$ where C_j is the circle $|z| = r_j$ oriented counter clockwise; here $R_1 < r_1 < r_2 < R_2$. On any circle $|z| = r$, with $R_1 < r < R_2$, we have $*du = r \cdot \frac{\partial u}{\partial r} d\theta$. Hence we have

$$\log r_1 \int_{C_1} r_1 \cdot \frac{\partial u}{\partial r} d\theta - \int_{C_1} u \, d\theta = \log r_2 \int_{C_2} r_2 \cdot \frac{\partial u}{\partial r} d\theta - \int_{C_2} u \, d\theta$$

or

$$\log r \int_{|z|=r} r \cdot \frac{\partial u}{\partial r} d\theta - \int_{|z|=r} u \, d\theta = -B$$

is independent of r (it is constant).

Applying the same argument to the functions $u_1 = 1$ (constant function) and $u_2 = u$, we obtain that

$$\int_{|z|=r} r \cdot \frac{\partial u}{\partial r} d\theta = A$$

is constant over the annulus D and hence is zero (let $r = 0$) if u is harmonic in the disk $\{z \in \mathbb{C}; |z| < R_2\}$.

Thus for a function u harmonic in an annulus, the arithmetic mean over concentric circles $|z| = r$ is a linear function of $\log r$

$$\frac{1}{2\pi} \int_{|z|=r} u \, d\theta = A \log r + B \; ;$$

and if u is harmonic in a disk, then $A = 0$ and the arithmetic mean is constant. In the latter case $B = u(0)$ by continuity (the reader should know other proofs of this fact.)

Changing the origin to z_0 we see that if u is harmonic in the disk $\{z \in \mathbb{C}; |z - z_0| < R\}$, then for $0 < r < R$

$$u(z_0) = \frac{1}{2\pi} \int_0^{2\pi} u(z_0 + r \, e^{i\theta}) \, d\theta \; ;$$

this is the Mean Value Property (MVP) for harmonic functions that was already established in Corollary 9.7 as a consequence of the fact that real-valued harmonic functions are locally real parts of analytic functions. From it one also obtains the area Mean Value Property

$$u(z_0) = \frac{1}{2\pi i r^2} \iint_{|z-z_0| \leq r} u(z) \, dz \, d\bar{z}. \tag{9.10}$$

REMARK 9.17. If $u : S^1 \to S^1$ is a homeomorphism, then Pu is also a homeomorphism, from $\{z; |z| < 1\}$ to itself. This useful observation is

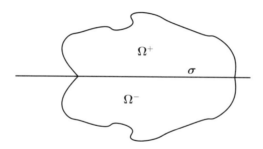

FIGURE 9.3. A symmetric region

not obvious at all. A generalization will be established in a companion volume on more advanced topics.

9.4. The Mean Value Property: a characterization

Harmonic functions satisfy the MVP, as we have seen in Corollary 9.7. As a matter of fact this property characterizes harmonic functions. The proof below is based on the solution to the Dirichlet problem.

THEOREM 9.18. *A continuous complex-valued function that satisfies the MVP is harmonic.*

PROOF. Let f be a continuous function on a domain D, let $\zeta \in D$, and let $r_0 > 0$ be sufficiently small so that $\operatorname{cl} U(\zeta, r_0) \subset D$ and f satisfies (5.6) for all $r \leq r_0$. It suffices to assume that f is real valued. Let v be the continuous function on $\{|z - \zeta| \leq r_0\}$ that is harmonic on $\{|z - \zeta| < r_0\}$ and agrees with f on $\{|z - \zeta| = r_0\}$. Then $f - v$ has the MVP in $\{|z - \zeta| < r_0\}$, and thus it attains its maximum and minimum on $\{|z - \zeta| = r_0\}$. Since $f = v$ on $\{|z - \zeta| = r_0\}$, we conclude that $f = v$ on $\{|z - \zeta| \leq r_0\}$ and thus that f is harmonic there. □

9.5. The reflection principle

We start with the simplest form of the reflection principle. Let Ω be a nonempty region in the complex plane that is *symmetric* about the real axis; that is, $\bar{z} \in \Omega$ if and only if $z \in \Omega$ (see Figure 9.3). Such a region must intersect the real axis nontrivially, and it is a disjoint union of three sets:

$$\Omega = \Omega^+ \cup \sigma \cup \Omega^-,$$

where

$$\Omega^+ = \{z \in \Omega;\ \Im z > 0\}, \quad \sigma = \Omega \cap \mathbb{R}, \quad \text{and} \quad \Omega^- = \{z \in \Omega;\ \Im z < 0\}.$$

REMARK 9.19. A function $z \mapsto f(z)$ on Ω is harmonic (analytic) if and only if the function $z \mapsto \overline{f(\bar{z})}$ is (see Exercise 9.2).

We concentrate on the holomorphic case. Assume that $f \in \mathbf{H}(\Omega)$ and f is real on at least one segment of σ; then $f(z) = \overline{f(\bar{z})}$ for all $z \in \Omega$.

PROOF. The function $z \mapsto g(z) = f(z) - \overline{f(\bar{z})}$ is analytic on Ω and vanishes on a subset of Ω with a limit point in Ω; it is thus identically zero on Ω.   □

The same conclusion holds if we merely assume that $f \in \mathbf{C}(\Omega^+ \cup \sigma)$, is analytic on Ω^+ and real on σ, since in this case the extension of f to Ω defined by $f(z) = \overline{f(\bar{z})}$ for $z \in \Omega^-$ satisfies the previous hypothesis. We can strengthen this statement considerably.

THEOREM 9.20. *Let Ω be a nonempty region in the complex plane that is symmetric about the real axis.*

If v is a real-valued, continuous function on $\Omega^+ \cup \sigma$, and it is harmonic on Ω^+ and zero on σ, then v has a harmonic extension to Ω that satisfies the symmetry condition $v(z) = -v(\bar{z})$.

Moreover, if v is the imaginary part of an analytic function f in $\mathbf{H}(\Omega^+)$, then f has an analytic extension to Ω that satisfies the symmetry condition $f(z) = \overline{f(\bar{z})}$.

PROOF. We use the symmetry to extend v to all of Ω. The resulting extension (also called v) is continuous on Ω, harmonic on $\Omega^+ \cup \Omega^-$, and vanishes on σ.

The issue of harmonicity of v in Ω is local; therefore we only need to show that v is harmonic in a neighborhood of each point $x \in \sigma$. For this purpose, consider an open disk D with center at x whose closure is contained in Ω. Let V be the unique function that is continuous on cl D, harmonic on its interior, and agrees with v on the boundary of D. Since v restricted to ∂D satisfies the symmetry condition $v(z) = -v(\bar{z})$, so does the function V (on cl D). Hence V vanishes on cl $D^+ \cap \mathbb{R}$. The function $V - v$ is continuous on cl D^+, harmonic on D^+, and vanishes on its boundary; hence it is identically zero on cl D^+. Similarly, $V - v = 0$ on cl D^-. We conclude that $V = v$ on D, and we have thus shown that v is harmonic on Ω.

The function f has a symmetric extension to $\Omega^+ \cup \Omega^-$ [that satisfies $f(z) = \overline{f(\bar{z})}$]. We only know that its imaginary part can be extended to all of Ω (and the extension vanishes on σ). We must use information on the imaginary part of f to draw conclusions about its real part. Again, the problem is local and we work with the disk D defined above. The real-valued harmonic function v on D has a harmonic conjugate $-u$ on this disk. The fact that harmonic conjugates are unique up to the addition of real constants allows us to normalize so that $u = \Re f$ in D^+. We study the function

$$U(z) = u(z) - u(\bar{z}), \quad z \in D.$$

This function vanishes on $D \cap \sigma$; hence,

$$\frac{\partial U}{\partial x} = 0 \quad \text{for} \quad (x, 0) \in D \cap \sigma.$$

Also, from the definition of U and the CR equations for the analytic function $u + \imath v$ on D,

$$\frac{\partial U}{\partial y} = 2\frac{\partial u}{\partial y} = -2\frac{\partial v}{\partial x} = 0 \quad \text{for} \quad (x, 0) \in D.$$

Thus the analytic function $2U_z = U_x - \imath U_y$ vanishes on $D \cap \sigma$ and is hence identically zero on D. Hence U is constant on D. Since it vanishes on $D \cap \sigma$, the constant must be zero. We have shown that on $D^+ \cup D^-$, the functions f and $u + \imath v$ agree. Since $u + \imath v$ is analytic on all of D, so is f.   □

Exercises

9.1. Prove that the equivalent forms for the Laplacian given in equations (9.2) and (9.3) are correct.

9.2. Show that a function $z \mapsto f(z)$ on Ω is harmonic (analytic) if and only if the function $z \mapsto \overline{f(\bar{z})}$ is.

9.3. Show that if $\omega = f\, dz + g\, d\bar{z}$ is a continuous differential form on a domain D, then

$$^*\omega = -\imath\,(f\, dz - g\, d\bar{z}).$$

9.4. We have shown that if u_1 and u_2 are real-valued harmonic functions on D, then

$$u_1\, {}^*du_2 - u_2\, {}^*du_1$$

is a closed form on D, and we have asserted that it also holds for complex-valued harmonic functions. Prove this assertion.

9.5. Prove the Maximum and Minimum Principles for real-valued harmonic functions

(1) as a general result for real-valued functions that satisfy the MVP, and then once again

(2) as a consequence of Harnack's inequalities for positive harmonic functions.

9.6. Let K be a compact subset of a domain $D \subseteq \mathbb{C}$, and let u be a positive harmonic function on D.

Show that there exists a constant $c \geq 1$ that depends only on K and D, but not on u, such that

$$\frac{1}{c} \leq \frac{u(z_1)}{u(z_2)} \leq c,$$

for all z_1 and $z_2 \in K$.

9.7. Let f be a continuous real-valued function on a domain D. Suppose that the partial derivatives $\frac{\partial^2 f}{\partial x^2}$ and $\frac{\partial^2 f}{\partial y^2}$ exist, and satisfy Laplace's equation $\triangle f = 0$. Show that f is harmonic on D.
Hint: Use the notation in the proof of Theorem 9.18. Let $\zeta = \xi + i\eta \in D$. Show first that for all $\epsilon > 0$, the function $F(z) = u(z) - v(z) + \epsilon(x - \xi)^2$ satisfies the maximum principle in $\{|z - \zeta| \leq r_0\}$.

9.8. Does the area Mean Value Property imply harmonicity for continuous functions?

9.9. If u is real valued and harmonic on $|z| < 1$, continuous on $|z| \leq 1$, and $u(e^{i\theta}) = \cos 2\theta + \sin 2\theta$, find $u\left(\frac{i}{2}\right)$.

9.10. Suppose that $u(0) = 1$, where u is harmonic and positive in a neighborhood of $\{z \in \mathbb{C}; |z| \leq 1\}$. Prove that $\frac{1}{7} \leq u\left(\frac{3}{4}\right) \leq 7$.

9.11. Let α be a real number. For $\zeta = e^{i\theta}$ with $\theta \in \mathbb{R}$, let

$$\varphi(\zeta) = \cos\theta + i\alpha\sin\theta.$$

Which of the following assertions are true for all α in $\mathbb{R}$? Which are true for some values of α?

a) The function $f(z) = \dfrac{1}{2\pi i} \displaystyle\int_{|\zeta|=1} \dfrac{\varphi(\zeta)}{\zeta - z}\, d\zeta$ is holomorphic for $|z| < 1$.

b) There exists a function f holomorphic for $|z| < 1$, continuous for $|z| \leq 1$, and satisfying $f(\zeta) = \varphi(\zeta)$ for $|\zeta| = 1$.

c) There exists a function f holomorphic for $|z| < 1$ such that $\Re f$ is continuous for $|z| \leq 1$ and satisfies $\Re f(\zeta) = \Re\varphi(\zeta)$ for $|\zeta| = 1$.

9.12. Let g be a continuous complex-valued function defined on S^1. Prove that there exists a continuous function f on $\{z \in \mathbb{C};\ |z| \leq 1\}$, such that f is holomorphic on $\{z \in \mathbb{C};\ |z| < 1\}$, and satisfies $f\big|_{S^1} = g$, if and only if

$$\int_{|\zeta|=1} g(\zeta)\,\zeta^n\, d\zeta = 0 \quad \text{for } n = 0, 1, 2, \dots .$$

9.13. Does there exist a function f holomorphic on $|z| < 1$ such that

$$\lim_{z \to \zeta} f(z) = \zeta + \zeta^{-1} \quad \text{for all } \zeta \text{ with } |\zeta| = 1?$$

9.14. Let $f \in \mathbf{C}^2(D)$. Show that f is holomorphic or anti-holomorphic on D if and only if f and f^2 (the square of f) are harmonic on D.

CHAPTER 10

Zeros of Holomorphic Functions

There are certain (classical families of) functions of a complex variable that mathematicians have studied frequently enough for them to acquire their own names. These functions are, of course, ones that develop naturally and repeatedly in various mathematical settings. Examples of such *named* functions include Euler's Γ-function, the Riemann ζ-function, and the Euler Φ-function. We will study only the first of these functions. There is a long history of synergy between the understanding of such functions and the development of complex analysis. Indeed, motivation for much of the theory and techniques of complex analysis was the desire to understand specific functions. In turn, the understanding of these functions has fed and continues to feed the development of the theory of complex variables.

Holomorphic functions in general and these classically studied functions in particular are often understood by their zeros. In this chapter we develop techniques to study the zeros of holomorphic functions. We show that one can always construct a meromorphic function with prescribed zeros and poles. To do so, we develop a theory for infinite products.

10.1. Infinite products

We begin with some language needed to discuss infinite products. We then develop lemmas that lead to Theorem 10.5, which relates the uniform convergence of certain infinite sums to the uniform convergence (to a holomorphic function) of corresponding infinite products.

DEFINITION 10.1. Let $u_n \in \mathbb{C}$ for each $n \in \mathbb{Z}_{>0}$, and set

$$p_n = (1 + u_1)(1 + u_2) \cdots (1 + u_n).$$

If $\lim_{n \to \infty} p_n$ exists and equals p, we write

$$p = \prod_{n=1}^{\infty} (1 + u_n).$$

We call p_n the *partial product* of the *infinite product* p. We say that the infinite product p *converges* if $\{p_n\}$ does.

LEMMA 10.2. *Let* $\{u_n\}_{n=1}^{\infty} \subset \mathbb{C}$ *and set*

$$p_N = \prod_{n=1}^{N}(1 + u_n) \quad and \quad p_N^* = \prod_{n=1}^{N}(1 + |u_n|).$$

Then

$$p_N^* \le e^{|u_1| + \cdots + |u_N|} \quad and \quad |p_N - 1| \le p_N^* - 1.$$

PROOF. We know that $x > 0$ implies that $e^x \ge 1 + x$. Therefore, $1 + |u_n| \le e^{|u_n|}$, so that $p_N^* \le e^{|u_1| + \cdots + |u_N|}$.

The second statement is proved by induction on N. For $N = 1$, $|p_1 - 1| = |u_1| = p_1^* - 1$.

For $N \ge 1$,

$$\begin{aligned}
|p_{N+1} - 1| &= |p_N(1 + u_{N+1}) - 1| \\
&= |(p_N - 1)(1 + u_{N+1}) + u_{N+1}| \\
&\le |p_N - 1| \cdot |1 + u_{N+1}| + |u_{N+1}| \,,
\end{aligned}$$

and by induction, this expression is

$$\begin{aligned}
&\le (p_N^* - 1) \cdot (1 + |u_{N+1}|) + |u_{N+1}| \\
&= p_{N+1}^* - 1 \,.
\end{aligned}$$

$\square$

THEOREM 10.3. *Suppose* $\{u_n\}$ *is a sequence of bounded functions on a set* S *such that* $\sum |u_n|$ *converges uniformly on* S.

Then the following hold:

(1) $f(z) = \displaystyle\prod_{n=1}^{\infty}(1 + u_n(z))$ *converges uniformly on* S.

(2) If $J : \mathbb{Z}_{>0} \to \mathbb{Z}_{>0}$ *is any bijection, then*

$$f(z) = \prod_{k=1}^{\infty}(1 + u_{J(k)}(z)).$$

(3) $f(z_0) = 0$ *if and only if* $u_n(z_0) = -1$ *for some* $n \in \mathbb{Z}_{>0}$.

PROOF. By uniform convergence of $\sum |u_n|$ on S, there exists $c \in \mathbb{R}_{>0}$ such that

$$\sup_{z \in S} \sum |u_n(z)| \le c.$$

Let

$$p_N(z) = \prod_{n=1}^{N}(1 + u_n(z)) \quad \text{and} \quad q_M(z) = \prod_{k=1}^{M}(1 + u_{J(k)}(z)).$$

We know that

$$|p_N| \le |p_N - 1| + 1 \le p_N^* \le e^{|u_1| + \cdots + |u_N|} \le e^c.$$

Choose ϵ with $0 < \epsilon < \frac{1}{4}$. Then there exists an $N_0 \in \mathbb{Z}_{>0}$ such that

$$\sum_{n=N_0}^{\infty} |u_n(z)| < \epsilon \quad \text{for all } z \in S;$$

in particular, $|u_n(z)| < \epsilon < \frac{1}{4}$ for all $z \in S$ and all $n > N_0$. Choose M_0 such that

$$\{1, 2, \ldots, N_0\} \subset \{J(1), J(2), \ldots, J(M_0)\}.$$

If $M, N > \max\{M_0, N_0\}$, then we can write

$$q_M(z) - p_N(z) = p_N(z)\left(\frac{\prod_1(1 + u_n(z))}{\prod_2(1 + u_n(z))} - 1\right).$$

Here the symbols $\prod_1$ and $\prod_2$ denote products taken over appropriate disjoint indices. For our purposes the important facts about $\prod_1$ and $\prod_2$ are that only indices $n > N_0$ appear, and that the indices that appear in the two products are disjoint.

Let us define $\widetilde{u}_n(z)$ by

$$1 + \widetilde{u}_n(z) = \begin{cases} 1 + u_n(z) & \text{if } n \text{ appears in } \prod_1 \\ \dfrac{1}{1 + u_n(z)} & \text{if } n \text{ appears in } \prod_2 \end{cases},$$

and let I be the union of the indexing sets in $\prod_1$ and $\prod_2$. Note that $I \subset \mathbb{Z}_{>0}$.

Now if a and $b \in \mathbb{C}$, $\delta \in \mathbb{R}_{>0}$, $|a| < \delta < \frac{1}{2}$, and $\dfrac{1}{1+a} = 1 + b$, then $|b| < 2\delta$. Therefore

$$|q_M(z) - p_N(z)| = |p_N(z)|\left|\prod_{n \in I}(1 + \widetilde{u}_n(z)) - 1\right|$$

$$\le |p_N(z)|\left(\prod_{n \in I}(1 + |\widetilde{u}_n(z)|) - 1\right) \le |p_N(z)|\left(\exp\left(\sum_{n \in I}|\widetilde{u}_n(z)|\right) - 1\right)$$

$$\le |p_N(z)|\left(\exp\left(\sum_{n=N_0}^{\infty}|\widetilde{u}_n(z)|\right) - 1\right) \le |p_N|\left(e^{2\epsilon} - 1\right).$$

We now claim that $e^x - 1 \leq 2x$ for $0 \leq x \leq \frac{1}{2}$. This claim may be verified as follows. Define $F(x) = e^x - 1 - 2x$ for $x \in \mathbb{R}$. Observe that $F(0) = 0$ and $F'(x) = e^x - 2$. Since the (real) exponential function is increasing, $F' \leq e^{\frac{1}{2}} - 2$ on $[0, \frac{1}{2}]$. But

$$e^{\frac{1}{2}} = 1 + \frac{1}{2} + \frac{(\frac{1}{2})^2}{2!} + \frac{(\frac{1}{2})^3}{3!} + \cdots$$

$$< 1 + \frac{1}{2} + \left(\frac{1}{2}\right)^2 + \cdots = \frac{1}{1 - \frac{1}{2}} = 2,$$

and thus F is nonpositive on $[0, \frac{1}{2}]$. Therefore

$$|q_M(z) - p_N(z)| \leq e^c 4\epsilon \quad \text{for all } z \in S,$$

and we can conclude as follows:

(1) If we let J be the identity map, then $p_N(z) \to f(z)$ uniformly for $z \in S$.

(2) For arbitrary J, we conclude that $q_M(z) \to f(z)$ uniformly in z.

(3) Since $p_{N_0} = p_M - (p_M - p_{N_0})$, we have, for sufficiently large M,

$$|p_{N_0}| \leq |p_M| + |p_M - p_{N_0}| \leq |p_M| + (e^{2\epsilon} - 1)|p_{N_0}|$$
$$\leq |p_M| + 4\epsilon |p_{N_0}|;$$

or, equivalently, that

$$|p_M(z)| \geq (1 - 4\epsilon)|p_{N_0}(z)|.$$

Therefore $f(z) = 0$ if and only if $p_{N_0}(z) = 0$. $\square$

THEOREM 10.4. *Assume* $0 \leq u_n < 1$.

(1) If $\displaystyle\sum_{n=1}^{\infty} u_n < \infty$, *then* $\displaystyle 0 < \prod_{n=1}^{\infty}(1 - u_n) < \infty$.

(2) If $\displaystyle\sum_{n=1}^{\infty} u_n = +\infty$, *then* $\displaystyle\prod_{n=1}^{\infty}(1 - u_n) = 0$.

PROOF. The first claim is a consequence of the previous theorem. To prove the second claim, we start with the observation that

$$1 - x \leq e^{-x} \quad \text{for } 0 \leq x \leq 1.$$

Let $p_N = (1-u_1)(1-u_2)\cdots(1-u_N)$. Since $p_1 \geq p_2 \geq \ldots \geq p_N \geq 0$, $\lim\limits_{N\to\infty} p_N = $ exists. We call it p. Now

$$0 \leq p \leq p_N = \prod_{n=1}^{N}(1-u_n) \leq e^{-(u_1+\cdots+u_N)}.$$

Since $\lim\limits_{N\to\infty} e^{-(u_1+\cdots+u_N)} = 0$, the theorem follows. $\qquad\square$

Finally, we can establish when an infinite product is holomorphic.

THEOREM 10.5. *Let D be a domain in $\mathbb{C}$, and suppose that $\{f_n\}$ is a sequence in $\mathbf{H}(D)$ with f_n not identically 0, and such that $\sum\limits_{n=1}^{\infty} |1 - f_n|$ converges uniformly on compact subsets of D.*

Then $\prod\limits_{n=1}^{\infty} f_n$ converges uniformly on compact subsets of D to a function $f \in \mathbf{H}(D)$, and

$$\nu_z(f) = \sum_{n=1}^{\infty} \nu_z(f_n) \quad \text{for all } z \in D.$$

PROOF. We only have to verify the last formula. We note that the sum in that formula is finite (that is, all but finitely many summands are zero). Let $z_0 \in D$, and let $K \subset D$ denote a compact set containing a neighborhood of z_0. There is an N in $\mathbb{Z}_{>0}$ such that $|1 - f_n(z)| < \frac{1}{2}$ for all $z \in K$ and all $n \geq N$. Therefore, $f_n(z) \neq 0$ for all $z \in K$ and for all $n \geq N$. Thus

$$\nu_{z_0}(f) = \nu_{z_0}\left(\prod_{n=1}^{N-1} f_n\right) + \nu_{z_0}\left(\prod_{n=N}^{\infty} f_n\right) = \sum_{n=1}^{N-1} \nu_{z_0}(f_n) + 0.$$

$\qquad\square$

10.2. Holomorphic functions with prescribed zeros

Our goal is to construct a holomorphic function with arbitrarily prescribed zeros (at a discrete set). To this end, we begin by defining *the elementary functions* first introduced by Weierstrass. We investigate some of their properties and then use them along with Theorem 10.5 to construct the required holomorphic function.

DEFINITION 10.6. Let $z \in \mathbb{C}$, and set

$$E_0(z) = 1 - z,$$

and, for $p \in \mathbb{Z}_{>0}$,

$$E_p(z) = (1-z)\exp\left(z + \frac{z^2}{2} + \cdots + \frac{z^p}{p}\right).$$

Note that, for all nonnegative integers p, $E_p(0) = 1$, and $E_p(z) = 0$ if and only if $z = 1$. Furthermore, the unique zero of E_p is simple.

LEMMA 10.7. *If* $|z| \leq 1$, *then* $|1 - E_p(z)| \leq |z|^{p+1}$ *for all nonnegative integers* p.

PROOF. The statement is clearly true if $p = 0$.
If $p \geq 1$, we have

$$E_p'(z) = (1-z)e^{z+\frac{z^2}{2}+\cdots+\frac{z^p}{p}}[1 + z + \cdots + z^{p-1}] - e^{z+\frac{z^2}{2}+\cdots+\frac{z^p}{p}}$$

$$= -z^p e^{z+\frac{z^2}{2}+\cdots+\frac{z^p}{p}}.$$

We therefore conclude that $\nu_0(-E_p') = p$. Further,

$$-E_p'(z) = z^p e^{z+\frac{z^2}{2}+\cdots+\frac{z^p}{p}}$$

$$= z^p \sum_{n=0}^{\infty} \frac{1}{n!}\left(z + \frac{z^2}{2} + \cdots + \frac{z^p}{p}\right)^n$$

$$= \sum_{n \geq p} b_n z^n,$$

with $b_p = 1$ and $b_n > 0$ for all $n \geq p$. Therefore

$$1 - E_p(z) = \sum_{n \geq p} \frac{b_n}{n+1} z^{n+1}.$$

Set

$$\phi(z) = \frac{1 - E_p(z)}{z^{p+1}}.$$

We observe that $\phi \in \mathbf{H}(\mathbb{C})$ and $\phi(z) = \sum_{n \geq 0} a_n z^n$, with $a_n > 0$ for all $n \in \mathbb{Z}_{\geq 0}$. For $|z| \leq 1$, we have

$$|\phi(z)| = \left|\sum_{n \geq 0} a_n z^n\right| \leq \sum_{n \geq 0} a_n |z^n| \leq \sum_{n \geq 0} a_n = \phi(1) = 1;$$

thus $|1 - E_p(z)| \leq |z|^{p+1}$ for $|z| \leq 1$. $\square$

THEOREM 10.8. *Assume that $\{z_n\}_{n=1}^{\infty}$ is a sequence of nonzero complex numbers with $\lim\limits_{n\to\infty} |z_n| = \infty$.*

Let $\{p_n\} \subseteq \mathbb{Z}_{\geq 0}$ be a sequence of nonnegative integers with the property that for all positive real numbers r, we have

$$\sum_{n=1}^{\infty} \left(\frac{r}{|z_n|} \right)^{1+p_n} < \infty. \tag{10.1}$$

Then the infinite product

$$P(z) = \prod_{n=1}^{\infty} E_{p_n}\left(\frac{z}{z_n} \right)$$

defines an entire function whose zero set is $\{z_1, z_2, \ldots\}$. More precisely, if $z = a$ appears $\nu \geq 0$ times in the above set of zeros, then $\nu_a(P) = \nu$.

Condition (10.1) is always satisfied for $p_n = n - 1$. Thus any discrete set in $\mathbb{C}$ is the zero set of an entire function.

PROOF. We first show that (10.1) holds for $p_n = n-1$. In this case we have to show convergence of the series $\sum a_n$, with $a_n = \left(\dfrac{r}{|z_n|} \right)^n$. But $|a_n|^{\frac{1}{n}} \to 0$ as $n \to \infty$, and the root test allows us to conclude that

$$\sum_{n=1}^{\infty} \left(\frac{r}{|z_n|} \right)^n < \infty.$$

Now let $\{p_n\}$ be any sequence of nonnegative integers satisfying condition (10.1); fix $r > 0$, and assume that $|z| \leq r$. From Lemma 10.7, we conclude that

$$\left| 1 - E_{p_n}\left(\frac{z}{z_n} \right) \right| \leq \left| \frac{z}{z_n} \right|^{p_n+1} \leq \left(\frac{r}{|z_n|} \right)^{p_n+1}.$$

Therefore we can apply Theorem 10.5 to conclude that $\prod E_{p_n}\left(\frac{z}{z_n} \right)$ converges uniformly on all compact subsets of $\mathbb{C}$ to an entire function that has the required zero set. $\square$

We next prove

THEOREM 10.9. *Let $f \in \mathbf{H}(\mathbb{C})$ have $\nu_0(f) = k$. Let $\{z_1, z_2, \ldots\}$ be the other zeros of f, listed according to their multiplicities in nondecreasing order.*

Then there exist a function $g \in \mathbf{H}(\mathbb{C})$ and a sequence of nonnegative integers $\{p_1, p_2, \ldots\}$ such that

$$f(z) = z^k e^{g(z)} \prod_{n=1}^{\infty} E_{p_n}\left(\frac{z}{z_n}\right).$$

PROOF. Choose any sequence of nonnegative integers $\{p_n\}$ such that (10.1) holds for all $r > 0$. Let

$$P(z) = \prod_{n=1}^{\infty} E_{p_n}\left(\frac{z}{z_n}\right) \quad \text{and} \quad G(z) = \frac{f(z)}{z^k P(z)}.$$

Then $G \in \mathbf{H}(\mathbb{C})$ and $G(z) \neq 0$ for all $z \in \mathbb{C}$. Since G is a nonvanishing entire function, there is a $g \in \mathbf{H}(\mathbb{C})$ with $e^{g(z)} = G(z)$ for all $z \in \mathbb{C}$. $\qquad \square$

THEOREM 10.10. *Let D be a proper subdomain of $\widehat{\mathbb{C}}$. Let A be a subset of D that has no limit point in D, and let ν be a function mapping A to $\mathbb{Z}_{>0}$.*

Then there exists a function $f \in \mathbf{H}(D)$ with $\nu_z(f) = \nu(z)$ for all $z \in A$, whose restriction to $D - A$ has no zeros.

PROOF. To begin, we make the following observations:

(1) A is either finite or countable. We let $|A|$ denote the cardinality of A.

(2) Without loss of generality, we may assume that $\infty \in D - A$ and that A is nonempty.

(3) If $|A| < \infty$, let $A = \{z_1, \ldots, z_n\}$ and let $\nu_j = \nu(z_j)$, for all $1 \leq j \leq n$. Choose $z_0 \in \widehat{\mathbb{C}} - D$ so that $z_0 \neq \infty$. We set in this case

$$f(z) = \frac{(z - z_1)^{\nu_1} \cdots (z - z_n)^{\nu_n}}{(z - z_0)^{\nu_1 + \cdots + \nu_n}},$$

and we note that f does not vanish on $\widehat{\mathbb{C}} - \{z_1, \ldots, z_n\}$ since $f(\infty) = 1$. We have thus established the theorem for finite sets A.

To prove the theorem for infinite sets A, let $K = \widehat{\mathbb{C}} - D$. Note that K is a compact subset of $\mathbb{C}$.

Let $\{a_n\}_{n=1}^{\infty}$ be a sequence whose terms consist of all $a \in A$, where each a is repeated $\nu(a)$ times.

We first claim that, for each positive integer n, we can choose a $\beta_n \in K$ such that $|\beta_n - a_n| \leq |\beta - a_n|$ for all $\beta \in K$. To see that this claim is valid, note that the function $z \mapsto l(z) = |z - a_n|$ for $z \in K$ is continuous on K and, therefore, achieves a minimum at some $\beta_n \in K$.

The function f we are seeking is

$$f(z) = \prod_{n=1}^{\infty} E_n \left(\frac{\alpha_n - \beta_n}{z - \beta_n} \right).$$

We show next that $\lim_{n \to \infty} |\beta_n - \alpha_n| = 0$. For if $|\beta_n - \alpha_n| \geq \delta$ for some $\delta > 0$ and infinitely many n, then for some subsequence $\{\alpha_{n_j}\}$ of $\{\alpha_n\}$,

$$\left| z - \alpha_{n_j} \right| \geq \delta \quad \text{for all } z \in K. \tag{10.2}$$

But a subsequence of this subsequence converges to some point $\alpha \in \widehat{\mathbb{C}}$. From (10.2), we conclude that $\alpha \notin K$. Thus we arrive at the contradiction that $\alpha \in D$ and is a limit point of A.

Next, we put $r_n = 2 |\alpha_n - \beta_n|$ and observe that the r_n converge to zero as n goes to infinity. We let K_0 be a compact subset of D, and we note that $r_n \to 0$ implies there is an $N \in \mathbb{Z}_{>0}$ such that $|z - \beta_n| > r_n$ for all $z \in K_0$ and all $n > N$. Since K and K_0 are disjoint compact subsets of $\widehat{\mathbb{C}}$, the distance δ_0 between them must be positive.

Thus

$$\left| \frac{\alpha_n - \beta_n}{z - \beta_n} \right| \leq \frac{r_n}{2r_n} = \frac{1}{2} \text{ for all } n > N \text{ and all } z \in K_0,$$

and hence

$$\left| 1 - E_n \left(\frac{\alpha_n - \beta_n}{z - \beta_n} \right) \right| \leq \left| \frac{\alpha_n - \beta_n}{z - \beta_n} \right|^{n+1} \leq \left(\frac{1}{2} \right)^{n+1}$$

for all $n > N$ and all $z \in K_0$.

Therefore the infinite product defining f converges. Now $f(z) = 0$ if and only if $E_n \left(\dfrac{\alpha_n - \beta_n}{z - \beta_n} \right) = 0$ for some $n \in \mathbb{Z}_{>0}$ if and only if $z = \alpha_n$ for some $n \in \mathbb{Z}_{>0}$. □

As an immediate corollary we obtain the following

THEOREM 10.11. *If D is a proper subdomain of $\widehat{\mathbb{C}}$, then $\mathbf{M}(D)$ is the field of fractions of the integral domain $\mathbf{H}(D)$.*

10.3. Euler's Γ-function

In this section we introduce an important function among whose remarkable properties is the fact that it extends the factorial function on the integers to an entire function. Our development is more brisk than in previous sections.

10.3.1. Basic properties. Define, for $z \in \mathbb{C}$,

$$G(z) = \prod_{n=1}^{\infty} \left(1 + \frac{z}{n}\right) e^{-\frac{z}{n}}.$$

The infinite product converges to an entire function with simple zeros at each negative integer.

We claim that

$$h(z) = z\, G(z)\, G(-z) = \frac{\sin \pi z}{\pi}, \quad \text{for all } z \in \mathbb{C}. \qquad (10.3)$$

Simple calculations show that

$$h(z) = z \prod_{n=1}^{\infty} \left(1 - \frac{z^2}{n^2}\right)$$

and hence, using (7.3) of Chapter 7,

$$\frac{h'(z)}{h(z)} = \frac{d}{dz} \log h(z) = \frac{1}{z} - \sum_{n=1}^{\infty} \frac{2z}{n^2 - z^2} = \pi \cot \pi z.$$

It follows that $h(z) = c \sin \pi z$ for some nonzero constant c. To evaluate c, we note that

$$\lim_{z \to 0} \frac{\sin \pi z}{z} = \pi, \quad \text{and} \quad \lim_{z \to 0} \frac{h(z)}{z} = 1,$$

and thus, we conclude that $c = \dfrac{1}{\pi}$.

The function $G(z - 1)$ is entire and has simple zeros at each non-positive integer and no other zeros. It follows that

$$G(z - 1) = z e^{\gamma(z)} G(z) \qquad (10.4)$$

for some entire function γ. We proceed to determine this function. Differentiating logarithmically both sides of the last equation, we obtain

$$\sum_{n=1}^{\infty} \left(\frac{1}{z - 1 + n} - \frac{1}{n}\right) = \frac{1}{z} + \gamma'(z) + \sum_{n=1}^{\infty} \left(\frac{1}{z + n} - \frac{1}{n}\right).$$

Since

$$\sum_{n=1}^{\infty} \left(\frac{1}{z - 1 + n} - \frac{1}{n}\right) = \sum_{n=0}^{\infty} \left(\frac{1}{z + n} - \frac{1}{n + 1}\right),$$

we conclude that

$$\gamma'(z) = \sum_{n=1}^{\infty} \left(\frac{1}{n} - \frac{1}{n + 1}\right) - 1 = 0.$$

Hence the function γ is constant. It is known as *Euler's constant*.

Returning to our function G, we observe that if we set $z = 1$ in (10.4), we obtain that $1 = G(0) = e^\gamma G(1)$. We determine γ as follows: From (10.4),

$$e^{-\gamma} = \prod_{n=1}^{\infty} \left(1 + \frac{1}{n}\right) e^{-\frac{1}{n}}.$$

Thus

$$e^{-\gamma} = \lim_{n\to\infty} (n+1)e^{-(1+\frac{1}{2}+\cdots+\frac{1}{n})},$$

and hence,

$$-\gamma = \lim_{n\to\infty} \left[\log(n+1) - \left(1 + \frac{1}{2} + \cdots + \frac{1}{n}\right)\right].$$

Since $\lim_{n\to\infty}(\log(n+1) - \log n) = \lim_{n\to\infty} \log(1 + \frac{1}{n}) = 0$, we obtain

$$\gamma = \lim_{n\to\infty} \left[\left(1 + \frac{1}{2} + \cdots + \frac{1}{n}\right) - \log n\right]. \tag{10.5}$$

Next set $H(z) = e^{\gamma z} G(z)$ and compute that

$$H(z-1) = e^{\gamma z} e^{-\gamma} G(z-1) = e^{\gamma z} z G(z) = z H(z).$$

We can now introduce

DEFINITION 10.12. *Euler's Γ-function* is defined by

$$\Gamma(z) = \frac{1}{zH(z)}, \quad \text{for } z \in \mathbb{C}.$$

Note that Γ is a meromorphic function on $\mathbb{C}$ with simple poles at $z = 0, -1, -2, \ldots$, and that it has no zeros.

The Γ-function satisfies a number of useful functional equations. We now derive some of these, which will lead up to (10.11), known as *Legendre's duplication formula*.

We start with

$$\Gamma(z) = \frac{e^{-\gamma z}}{z} \prod_{n=1}^{\infty} \left(1 + \frac{z}{n}\right)^{-1} \cdot e^{\frac{z}{n}}. \tag{10.6}$$

The Γ-function satisfies the functional equation

$$\Gamma(z+1) = \frac{1}{(z+1)H(z+1)} = \frac{1}{H(z)} = z\Gamma(z). \tag{10.7}$$

Furthermore, it follows from (10.3) that

$$\Gamma(z)\Gamma(1-z) = \frac{\pi}{\sin \pi z}. \tag{10.8}$$

A simple calculation shows that

$$\Gamma(1) = e^{-\gamma} \prod_{n=1}^{\infty} \left(1 + \frac{1}{n}\right)^{-1} e^{\frac{1}{n}} = 1.$$

Together with (10.7), this implies that

$$\Gamma(n) = (n-1)!, \quad \text{for all } n \in \mathbb{Z}_{>0}.$$

Also, $(\Gamma\left(\frac{1}{2}\right))^2 = \dfrac{\pi}{\sin\frac{\pi}{2}} = \pi$ implies that

$$\Gamma\left(\frac{1}{2}\right) = \sqrt{\pi}.$$

We derive some other properties of Euler's Γ-function that we will need. We start with a calculation, from (10.6):

$$\frac{d}{dz}\frac{\Gamma'(z)}{\Gamma(z)} = \frac{d}{dz}\frac{d}{dz}(\log(\Gamma(z)))$$

$$= \frac{d}{dz}\left(-\gamma - \frac{1}{z} - \sum_{n=1}^{\infty}\left(\frac{1}{z+n} - \frac{1}{n}\right)\right) = \sum_{n=0}^{\infty}\left(\frac{1}{z+n}\right)^2. \quad (10.9)$$

Both functions

$$z \mapsto \Gamma(2z) \quad \text{and} \quad z \mapsto \Gamma(z)\Gamma\left(z + \frac{1}{2}\right)$$

have simple poles precisely at the points $0, -1, -2, \ldots$ and $-\frac{1}{2}, -\frac{3}{2}, \ldots$. The ratio of the two functions is hence entire without zeros. The next calculation will show more:

$$\frac{d}{dz}\left(\frac{\Gamma'(z)}{\Gamma(z)}\right) + \frac{d}{dz}\left(\frac{\Gamma'(z+\frac{1}{2})}{\Gamma(z+\frac{1}{2})}\right) = \sum_{n=0}^{\infty}\frac{1}{(z+n)^2} + \sum_{n=0}^{\infty}\frac{1}{(z+n+\frac{1}{2})^2}$$

$$= 4\left(\sum_{n=0}^{\infty}\frac{1}{(2z+2n)^2} + \sum_{n=0}^{\infty}\frac{1}{(2z+2n+1)^2}\right)$$

$$= 4\left(\sum_{m=0}^{\infty}\frac{1}{(2z+m)^2}\right) = 2\frac{d}{dz}\left(\frac{\Gamma'(2z)}{\Gamma(2z)}\right).$$

Therefore, for some constant a, we have

$$2\frac{\Gamma'(2z)}{\Gamma(2z)} = \frac{\Gamma'(z)}{\Gamma(z)} + \frac{\Gamma'(z+\frac{1}{2})}{\Gamma(z+\frac{1}{2})} - a,$$

or, equivalently,

$$\frac{d}{dz}\log\Gamma(2z) = \frac{d}{dz}\log\Gamma(z)\Gamma\left(z + \frac{1}{2}\right) - a.$$

Thus, for some constant b, we have

$$\operatorname{Log} \Gamma(2z) = \operatorname{Log} \Gamma(z)\Gamma\left(z + \frac{1}{2}\right) - az - b$$

or, equivalently,

$$\Gamma(2z)e^{az+b} = \Gamma(z)\Gamma\left(z + \frac{1}{2}\right). \tag{10.10}$$

Next work backward to determine a and b. Setting $z = \frac{1}{2}$ in (10.10), we obtain $1 \cdot e^{\frac{1}{2}a+b} = \sqrt{\pi}$; that is, $\frac{1}{2}a + b = \frac{1}{2}\log \pi$. Setting $z = 1$ in (10.10), we obtain $e^{a+b} = \frac{1}{2}\sqrt{\pi}$; that is, $a + b = \frac{1}{2}\log \pi - \log 2$.

Thus $a = -2\log 2$ and $b = \frac{1}{2}(\log \pi) + \log 2$, and we have established

Legendre's duplication formula

$$\sqrt{\pi}\Gamma(2z) = 2^{2z-1}\Gamma(z)\Gamma\left(z + \frac{1}{2}\right). \tag{10.11}$$

10.3.2. Estimates for $\Gamma(z)$. The estimate of $\Gamma(z)$ for large values of $|z|$ that is found in this section is known as **Stirling's formula**.

To derive this formula, we first express the partial sums $\sum_{k=0}^{n}\left(\frac{1}{z+k}\right)^2$ of $\frac{d}{dz}\frac{\Gamma'(z)}{\Gamma(z)}$ [see (10.9)] as a *convenient* line integral.

View $z = x + \iota y$ as a (fixed) parameter and $\zeta = \xi + \iota \eta$ as a variable, and define

$$\Phi(\zeta) = \frac{\pi \cot \pi \zeta}{(z + \zeta)^2}, \quad \text{for } \zeta \in \mathbb{C}.$$

The function Φ has singularities at $\zeta = -z$ and at $\zeta \in \mathbb{Z}$; if $z \notin \mathbb{Z}$, a double pole at $-z$ and simple poles at the integers. Let Y be a positive real number, n be a nonnegative integer, and K be the rectangle in the ζ-plane described by $-Y \leq \eta \leq Y$ and $0 \leq \xi \leq n + \frac{1}{2}$ (see Figure 10.1).

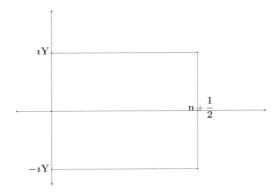

FIGURE 10.1. The rectangle K

Then the Residue Theorem yields

LEMMA 10.13. *For* $z \notin \mathbb{Z}_{\geq 0}$,

$$pr.\ v.\ \frac{1}{2\pi i} \int_{\partial K} \Phi(\zeta)\,d\zeta = -\frac{1}{2z^2} + \sum_{\nu=0}^{n} \frac{1}{(z+\nu)^2}.$$

We plan to let $Y \to \infty$ and $n \to \infty$. We thus have to study several line integrals, as follows.

10.3.2.1. *The integral over the horizontal sides:* $\eta = \pm Y$. On the horizontal sides $\eta = \pm Y$, $\cot \pi\zeta$ converges uniformly to $\pm i$ as Y goes to ∞. Thus $\dfrac{\cot \pi\zeta}{(z+\zeta)^2}$ converges to 0 on each of the line segments $\xi \geq 0, \eta = \pm Y$ as $Y \to \infty$. We need to show that

$$\lim_{n\to\infty} \lim_{Y\to\infty} \int_0^{n+\frac{1}{2}} \frac{\cot \pi(\xi \pm iY)}{(z + \xi + iY)^2}\,d\xi = 0.$$

Since we can control the speed with which Y and n approach infinity, this presents a small challenge; for example, we can set $Y = n^2$ and then let $n \to \infty$.

10.3.2.2. *The integral over the vertical side* $\xi = n + \frac{1}{2}$. On the vertical line $\xi = n+\frac{1}{2}$, $\cot \pi\zeta$ is bounded since $\cot$ is a periodic function. Thus we conclude that for some constant c,

$$\left| \int_{\xi=n+\frac{1}{2}} \Phi(\zeta)d\zeta \right| \leq c \int_{\xi=n+\frac{1}{2}} \frac{d\eta}{|\zeta + z|^2}.$$

On $\xi = n + \frac{1}{2}$, we have $\bar\zeta = n + \frac{1}{2} - i\eta = 2n + 1 - \zeta$. We then use residue calculus (as in Case (1) of Section 6.5) to conclude that

$$\frac{1}{i} \int_{\xi=n+\frac{1}{2}} \frac{d\zeta}{|\zeta + z|^2} = \frac{1}{i} \int_{\xi=n+\frac{1}{2}} \frac{d\zeta}{(\zeta + z)(2n + 1 - \zeta + \bar z)} = \frac{2\pi}{2n + 1 + 2x}.$$

Therefore,

$$\lim_{n\to\infty} \int_{\xi=n+\frac{1}{2}} \frac{d\eta}{|\zeta + z|^2} = 0.$$

10.3.2.3. *The integral over the imaginary axis.* We now turn to the computation of the principal value of the integral over the imaginary axis, which may be written as follows.

$$\frac{1}{2} \int_0^\infty \cot \pi i\eta \left[\frac{1}{(i\eta + z)^2} - \frac{1}{(i\eta - z)^2} \right] d\eta$$

$$= -\int_0^\infty \coth \pi\eta \frac{2\eta z}{(\eta^2 + z^2)^2} d\eta.$$

We now let Y and n tend to ∞ in Lemma 10.13 to conclude that

$$\frac{d}{dz}\frac{\Gamma'(z)}{\Gamma(z)} = \sum_{n=0}^{\infty} \frac{1}{(z+n)^2}$$

$$= \frac{1}{2z^2} + \int_{+\infty}^{-\infty} \Phi(\zeta)\, d\zeta = \frac{1}{2z^2} + \int_{0}^{\infty} \coth \pi\eta \frac{2\eta z}{(\eta^2 + z^2)^2}\, d\eta\,.$$

Replacing $\coth \pi\eta = 1 + \dfrac{2}{e^{2\pi\eta} - 1}$ in the above expression and noting that

$$\int_{0}^{\infty} \frac{2\eta z}{(\eta^2 + z^2)^2}\, d\eta = \frac{1}{z}\,,$$

we obtain

$$\frac{d}{dz}\frac{\Gamma'(z)}{\Gamma(z)} = \frac{1}{2z^2} + \int_{0}^{\infty} \left(1 + \frac{2}{e^{2\pi\eta} - 1}\right) \frac{2\eta z}{(\eta^2 + z^2)^2}\, d\eta$$

$$= \frac{1}{z} + \frac{1}{2z^2} + \int_{0}^{\infty} \frac{4\eta z}{(\eta^2 + z^2)^2} \frac{d\eta}{e^{2\pi\eta} - 1}\,.$$

We restrict z to $\Re z > 0$ and note that we can integrate under the integral sign with respect to z to conclude

$$\frac{\Gamma'(z)}{\Gamma(z)} = \widetilde{C} + \mathrm{Log}\, z - \frac{1}{2z} - \int_{0}^{\infty} \frac{2\eta}{\eta^2 + z^2} \frac{d\eta}{e^{2\pi\eta} - 1}\,.$$

Using integration by parts, we see that

$$-\int_{0}^{\infty} \frac{2\eta}{\eta^2 + z^2} \frac{d\eta}{e^{2\pi\eta} - 1} = \frac{1}{\pi} \int_{0}^{\infty} \frac{z^2 - \eta^2}{(\eta^2 + z^2)^2} \log(1 - e^{-2\pi\eta})\, d\eta\,;$$

therefore,

$$\frac{\Gamma'(z)}{\Gamma(z)} = \widetilde{C} + \mathrm{Log}\, z - \frac{1}{2z} + \frac{1}{\pi} \int_{0}^{\infty} \frac{z^2 - \eta^2}{(\eta^2 + z^2)^2} \log(1 - e^{-2\pi\eta})\, d\eta\,,$$

and we conclude that

$$\mathrm{Log}\, \Gamma(z) = C' + Cz + \left(z - \frac{1}{2}\right) \cdot \mathrm{Log}\, z + J(z), \qquad (10.12)$$

where

$$J(z) = \frac{1}{\pi} \int_{0}^{\infty} \frac{z}{(\eta^2 + z^2)^2} \log \frac{1}{1 - e^{-2\pi\eta}}\, d\eta.$$

If $z \to \infty$ and z stays away from $i\mathbb{R}$, then $J(z) \to 0$. We have almost established

THEOREM 10.14 (**Stirling's formula**). *For $\Re z > 0$,*

$$\Gamma(z) = \sqrt{2\pi}\, z^{z-\frac{1}{2}} e^{-z} e^{J(z)}. \qquad (10.13)$$

PROOF. We know from (10.12) that

$$\Gamma(z) = e^{C' + Cz} z^{z - \frac{1}{2}} e^{J(z)}, \tag{10.14}$$

and we only need to determine the constants C' and C, which we now do using (10.14) and the two functional equations (10.7) and (10.8) already derived for Γ.

Replacing $\Gamma(z)$ by the RHS of (10.14) (and also $\Gamma(z + 1)$ by the corresponding value) in $\Gamma(z + 1) = z\Gamma(z)$, we obtain

$$C = -(z + \frac{1}{2}) \operatorname{Log}\left(1 + \frac{1}{z}\right) + J(z) - J(z + 1),$$

and letting z tend to ∞, we conclude that $C = -1$.

To obtain C' one can proceed in a similar manner, replacing $\Gamma(z)$ and $\Gamma(1 - z)$ by the corresponding RHS of (10.14) in the equality $\Gamma(z)\Gamma(1 - z) = \dfrac{\pi}{\sin \pi z}$, with $z = \frac{1}{2} + iy$. We leave the details to the reader.   □

COROLLARY 10.15.

$$\lim_{n \to \infty} \frac{n!}{\sqrt{2\pi n} \left(\frac{n}{e}\right)^n} = 1.$$

PROOF. Note that

$$(n + 1)^{n + \frac{1}{2}} = n^{n + \frac{1}{2}} \left(1 + \frac{1}{n}\right)^n \left(1 + \frac{1}{n}\right)^{\frac{1}{2}},$$

and therefore,

$$\lim_{n \to \infty} \frac{(n + 1)^{n + \frac{1}{2}}}{e \, n^{n + \frac{1}{2}}} = 1. \tag{10.15}$$

Applying Stirling's formula (10.13) with $z = n + 1$, we obtain

$$\Gamma(n + 1) = \sqrt{2\pi}(n + 1)^{n + \frac{1}{2}} e^{-(n+1)} e^{J(n+1)}.$$

Since we already know that $\lim\limits_{n \to i\infty} J(n + 1) = 0$, the claim is proved.
                                                                  □

With some additional work we can prove the following integral expression for the Γ-function:

$$\Gamma(z) = \int_0^\infty e^{-t} \cdot t^{z-1} \, dt \,, \quad \text{for } \Re z > 0. \tag{10.16}$$

PROOF. Denote for the moment the RHS of (10.16) by $F(z)$. It certainly defines a holomorphic function whenever the integral converges; that is, in the right half-plane given by $\Re z > 0$. Since

$$F(z+1) = \int_0^\infty e^{-t} \cdot t^z dt = z \int_0^\infty e^{-t} \cdot t^{z-1} dt = zF(z) \quad \text{for } \Re z > 0,$$

we see that F and Γ satisfy the same functional equation. Thus

$$\frac{F(z+1)}{\Gamma(z+1)} = \frac{F(z)}{\Gamma(z)} \quad \text{for } \Re z > 0,$$

and the function F can be extended to be defined on all of $\mathbb{C}$. For z with $1 \le \Re z \le 2$,

$$|F(z)| \le \int_0^\infty e^{-t} \cdot t^{\Re z - 1} dt = F(\Re z),$$

and since F is a continuous function on the closed interval $[1, 2]$, it is bounded there. The last estimate shows that F is bounded on the strip $1 \le \Re z \le 2$. We need a lower bound for $|\Gamma|$ on the same strip; it suffices to determine how negative $\log |\Gamma| = \Re \log \Gamma$ can become. We do so as follows. From (10.12), we see that

$$\log |\Gamma(z)| = \frac{1}{2} \log 2\pi - \Re z + \left(\Re z - \frac{1}{2} \right) \log |z| - \Im z \operatorname{Arg} z + \Re J(z).$$

Only the term $-\Im z \operatorname{Arg} z$ approaches $\pm\infty$ as $\Im z$ approaches $+\infty$. This term is comparable with $-\dfrac{\pi}{2} |\Im z|$. Thus as $\Im z$ goes to $\pm\infty$, $\left| \dfrac{F}{\Gamma} \right|$ grows at most like a constant multiple of $\exp\left(-\dfrac{\pi}{2} |\Im z| \right)$.

Since $\dfrac{F}{\Gamma}$ is periodic, it is a function of $W = e^{2\pi i z}$ (on the punctured plane $\mathbb{C} - \{0\}$); it has an isolated singularity at $W = 0$. As W approaches 0, $\dfrac{F}{\Gamma}$ grows at most as $|W|^{-\frac{1}{2}}$, and as W approaches ∞, $\dfrac{F}{\Gamma}$ grows at most as $|W|^{\frac{1}{2}}$. Hence the singularities at both ends are removable and $\dfrac{F}{\Gamma}$ is constant. Since $F(1) = 1 = \Gamma(1)$, this constant is 1. $\qquad\square$

10.4. The field of meromorphic functions

It is convenient to introduce the following

DEFINITION 10.16. Let D be a plane domain. A *divisor* on D is either the empty set $\emptyset$ or a formal expression

$$\prod_i z_i^{\nu_i},$$

where the $\{z_i\}$ form a discrete set in D and $\nu_i \in \mathbb{Z}$. We will also write a divisor $\mathcal{D}$ as

$$\mathcal{D} = \prod_{z \in D} z^{\nu_z(\mathcal{D})},$$

with the understanding that $\nu_z(\mathcal{D}) \in \mathbb{Z}$ for all $z \in D$ and $\nu_z(\mathcal{D}) = 0$ for all z not in a discrete set (that depends on the divisor, of course) Note that $\nu_z(\emptyset) = 0$ for all $z \in D$.

There is a commutative law of multiplication for divisors, where the empty set is the unit element, and if $\mathcal{D}_1 = \prod_{z \in D} z^{\nu_z(\mathcal{D}_1)}$ and $\mathcal{D}_2 = \prod_{z \in D} z^{\nu_z(\mathcal{D}_2)}$ are two divisors, then

$$\mathcal{D}_1 \cdot \mathcal{D}_2 = \prod_{z \in D} z^{\nu_z(\mathcal{D}_1)+\nu_z(\mathcal{D}_2)} .$$

The set of all divisors on D with this operation becomes a commutative group, denoted by $\mathrm{Div}(D)$.

In particular, to a nonidentically zero meromorphic function f on D, we can associate its divisor (f) defined by

$$(f) = \prod_{z \in D} z^{\nu_z(f)},$$

where $\nu_z(f)$ denotes the order of the function f at the point z in D (see Section 3.5). In particular, to any constant (nonzero) function in D we associate the divisor given by the empty set. A divisor of a nonzero meromorphic function in D is called a *principal divisor*.

Note that the set of all principal divisors is a subgroup of $\mathrm{Div}(D)$, and that the function $(\cdot)$ that associates with each nonzero meromorphic function f on D its divisor (f) is a homomorphism from the multiplicative group $\mathbf{M}(D) - \{0\}$ to $\mathrm{Div}(D)$, whose image is the subgroup of principal divisors. Much more is true as shown next.

THEOREM 10.17. *Let D be a domain in $\mathbb{C}$. The map*

$$(\cdot) : \mathbf{M}(D) - \{0\} \to \mathrm{Div}(D)$$

is a surjective homomorphism, whose kernel is the set of nowhere vanishing holomorphic functions on D.

PROOF. Write $\mathcal{D} = \frac{\mathcal{D}_1}{\mathcal{D}_2}$, where $\mathcal{D}_1$ and $\mathcal{D}_2$ are relatively prime integral divisors (see Exercise 10.5) and use Theorem 10.10.   □

10.5. Infinite Blaschke products

Let

$$\mathcal{D} = \prod_i a_i^{\nu_i}$$

be an integral divisor (see Exercise 10.5) on the unit disk $\mathbb{D}$. It is clear that the definitions of Section 8.5 associate a Blaschke product $B_{\mathcal{D}}$ with the divisor $\mathcal{D}$. We end this book with the statement of two theorems. The proof of the first is left as a formal exercise, Exercise 10.10. The proof of the second theorem is left as food for thought. It requires material that we have not developed.

THEOREM 10.18. *Let $\mathcal{D} = \prod_i a_i^{\nu_i}$ be an integral divisor on the unit disk $\mathbb{D}$. The Blaschke product $B_{\mathcal{D}}$ converges to a bounded analytic function on $\mathbb{D}$ if and only if*

$$\sum_i \nu_i \left(1 - |a_i|\right) < \infty.$$

If $\sum_i \nu_i \left(1 - |a_i|\right) = \infty$, then the Blaschke product converges to the constant zero function on the disk.

THEOREM 10.19. *Let $\mathcal{D} = \prod_i a_i^{\nu_i}$ be an integral divisor on the unit disk $\mathbb{D}$. Then there exists a bounded analytic function on $\mathbb{D}$ with divisor $\mathcal{D}$ if and only if $\sum_i \nu_i \left(1 - |a_i|\right) < \infty$.*

Exercises

10.1. Show that $\prod_{n=2}^{\infty}(1 - \frac{1}{2^n}) = \frac{1}{2}$.

10.2. Let $\{a_n\}$ be a sequence in $\mathbb{Z}_{\neq -1}$. The infinite product $\prod_{n=1}^{\infty}(1 + a_n)$ is said to be *absolutely convergent* if the corresponding series $\sum_{n=1}^{\infty} \text{Log}(1 + a_n)$ converges absolutely.

(1) Show that $\prod_{n=1}^{\infty}(1 + a_n)$ converges absolutely if and only if $\sum_{n=0}^{\infty} |a_n|$ converges.

(2) Show that the value of an absolutely convergent product does not change if the factors are reordered.

(3) Find examples that show that the convergence of $\sum_{n=0}^{\infty} a_n$ is neither necessary nor sufficient for the convergence of $\prod_{n=1}^{\infty}(1 + a_n)$.

10.3. Show that $\Gamma(\frac{1}{6}) = 2^{-\imath}(\frac{3}{\pi})^{\frac{1}{2}} \cdot (\Gamma(\frac{1}{3}))^2$.

10.4. Find the residues of Γ at the poles $z = -n$, $n \geq 1$.

10.5. A divisor $\mathcal{D}$ on a domain D is *integral* if $\nu_z(\mathcal{D}) \geq 0$ for all $z \in D$. Define appropriately the *greatest common divisor* $\gcd(\mathcal{D}_1, \mathcal{D}_2)$ and the *least common multiple* $\text{lcm}(\mathcal{D}_1, \mathcal{D}_2)$ of two integral divisors $\mathcal{D}_1$ and $\mathcal{D}_2$. Show that both the gcd and the lcm exist, and obtain formulas for them. We say that the integral divisors $\mathcal{D}_1$ and $\mathcal{D}_2$ are *relatively prime* if their gcd is the empty set.

10.6. Show that two nonzero meromorphic functions f and g in D give rise to the same divisor if and only if there exists a function h in $\mathbf{H}(D)$ that does not vanish in D and such that $f = gh$.

10.7. Show that a principal divisor on D is integral if and only if it is the divisor of an analytic function on D.

10.8. Show that if f and g are analytic functions in D (where at least one is not the zero function), then there exists a function $h \in \mathbf{H}(D)$ such that h is a greatest common divisor for f and g. That is, h divides f and g (in $\mathbf{H}(D)$) and it is divisible by every holomorphic function dividing both f and g.

If f and g are not identically 0, show that

$$(h) = \gcd((f), (g)).$$

Hint: Apply Theorem 10.10.

10.9. Replace $\Gamma(z)$ and $\Gamma(1-z)$ by the corresponding RHS of (10.14) in $\Gamma(z)\Gamma(1-z) = \dfrac{\pi}{\sin \pi z}$, with $z = \frac{1}{2} + \imath y$ in the proof of Stirling's formula (Theorem 10.14) to obtain C' in a similar manner to that used to obtain C. Give full details.

10.10. Prove Theorem 10.18.

The following exercises lead to a proof of the fact that every finitely generated ideal in $\mathbf{H}(D)$ is a principal ideal.

10.11. Show that every nonempty collection of holomorphic functions in D, except for the set consisting of the single function zero, has a greatest common divisor.

10.12. Show that if h is a greatest common divisor for $f_1, \ldots, f_n$ in $\mathbf{H}(D)$, then there exist $g_1, \ldots, g_n$ in $\mathbf{H}(D)$ such that

$$f_1 g_1 + \ldots + f_n g_n = h.$$

Hint: First consider the case when $n = 2$ and $h = 1$.

10.13. Let $f_1, \ldots, f_n$ be holomorphic functions in D, and consider the ideal I generated by them in $\mathbf{H}(D)$:

$$I = \{f_1 \, g_1 + \ldots + f_n \, g_n \,;\, g_j \in \mathbf{H}(D)\} \,.$$

Prove that I is a principal ideal; that is, there exists a function f in $\mathbf{H}(D)$ such that

$$I = \{f \, g \,;\, g \in \mathbf{H}(D)\} \,.$$

10.14. Characterize the principal maximal ideals in the ring of holomorphic functions on a plane domain. Do the concepts of "principal maximal" and "maximal principal" ideals coincide?

BIBLIOGRAPHICAL NOTES

The references required for proofs and definitions were included in the body of the book. The purpose of these notes is to list a number of basic books on complex analysis that the authors of this volume have found useful and interesting. This is an incomplete list reflecting the tastes and limited knowledge of the authors. We list three categories of books.

1. **Undergraduate Texts:** The texts by Churchill and Brown [5], Derrick [7], and Silverman [25] are each very appropriate for an undergraduate course that is centered around applications; Palka [22] is a very thorough and careful undergraduate text that is also used frequently in graduate courses; Marsden [19] is more theoretical, whereas Bak–Newman [2] is the only undergraduate text on the subject that ends up with a treatment of the Prime Number Theorem.

2. **Graduate Texts:** Ahlfors [1] is among the outstanding mathematics books in any field. Another classic treatment is the book by Nevanlinna-Paatero [21]; whereas Narasimhan [20] is an outstanding modern treatment. Cartan [4] starts with a treatment of formal power series. The first of the two Hille volumes [10] is a standard introduction to the subject; the second [11] deals with many interesting special topics. The book by Heins [9] covers many prerequisites currently dealt with on other courses and some advanced topics. Conway's book [6] is very concise yet complete. It includes the big Picard theorem. Berenstein–Gay [3] is in tune with more recent, modern developments in complex variables. The Greene–Krantz text [8] is a treatment of complex variables as an outgrowth of real multivariate calculus.

   Another standard reference is Lang [18]. Lang and Bers probably influenced each other's views of complex variables as a result of long discussions in the Columbia mathematics department fifth floor lounge while Lang was writing his book and Bers was teaching Complex Variables I and II during the 1966–67 academic year.

The first chapter of Hörmander [12] serves as a wonderful review for those who have learned one variable complex analysis and are interested in seeing how it leads naturally to a study of several variables. Rudin [23] is an integrated treatment of real and complex analysis.

3. **Problem Books:** Some paperback books contain problems sets and their solutions. Classic is a five-volume set by Konrad Knopp. Whenever Bers taught the year-long graduate Complex Analysis course at Columbia, he taught without a text and simply told his students to read and work through all of the problems in Knopp's books on the theory of functions [13]–[17] (Bers translated one of these five volumes from the German [15]). Knopp's problem books are still eminently relevant and useful. A paperback by Rami Shakarchi [24] includes solutions to all of the undergraduate-level problems from Lang's graduate text (problems from the first eight chapters) and solutions to selected problems from the more advanced chapters.

Bibliography

[1] L. V. Ahlfors, *Complex Analysis (third edition)*, McGraw-Hill, 1979.

[2] J. Bak and D. J. Newman, *Complex Analysis*, Springer-Verlag, 1982.

[3] C. A. Berenstein and R. Gay, *Complex Variables, an Introduction*, Graduate Texts in Mathematics, vol. 125, Springer-Verlag, 1991.

[4] H. Cartan, *Elementary Theory of Analytic Functions of One or Several Complex Variables*, Addison-Wesley, 1963.

[5] R. V. Churchill and J. W. Brown, *Complex Analysis and Applications (fifth edition)*, McGraw-Hill, 1990.

[6] J. B. Conway, *Functions of One Complex Variable (second edition)*, Springer, 1978.

[7] William R. Derrick, *Complex Analysis and Applications (second edition)*, Wadsworth International Group, 1982.

[8] R. E. Greene and S. G. Krantz, *Function Theory of one Complex Variable*, John Wiley & Sons, Inc., 1997.

[9] M. Heins, *Complex Function Theory*, Acadeic Press, 1968.

[10] E. Hille, *Analytic Function Theory, Volume I*, Blaisdell, 1959.

[11] _____, *Analytic Function Theory, Volume II*, Blaisdell, 1962.

[12] L. Hörmander, *An Introduction to Complex Analysis in Several Variables*, Van Nostrand, 1966.

[13] K. Knopp, *Theory of Functions I. Elements of the General Theory of Analytic Functions*, Dover Publications, 1945.

[14] _____, *Theory of Functions II. Applications and Continuation of the General Theory*, Dover Publications, New York, 1947.

[15] _____, *Problem Book in the Theory of Functions. Volume 1. Problems in the Elementary Theory of Functions*, Dover Publications Inc., New York, N. Y., 1948, Translated by Lipman Bers.

[16] _____, *Elements of the Theory of Functions*, Dover Publications Inc., New York, 1953, Translated by Frederick Bagemihl. MR MR0051295 (14,458b)

[17] _____, *Problem Book in the Theory of Functions. Volume II. Problems in the Advanced Theory of Functions*, Dover Publications Inc., New York, N. Y., 1953, Translated by F. Bagemihl.

[18] S. Lang, *Complex Analysis*, fourth ed., Graduate Texts in Mathematics, vol. 103, Springer-Verlag, New York, 1999.

[19] J. E. Marsden, *Basic Complex Analysis*, W. H. Freeman and Company, 1973.

[20] R. Narasimhan, *Complex analysis in One Variable*, Birkhäuser Verlag, 1985.

[21] R. Nevanlinna and V. Paatero, *Introduction to Complex Analysis*, Addison-Wesley, 1964.

[22] B. Palka, *An Introduction to Complex Function Theory*, Springer-Verlag, New York, 1991.

[23] W. Rudin, *Real and Complex Analysis, Third Edition*, McGraw-Hill, 1987.

[24] R. Shakarchi, *Problems and Solutions for Complex Analysis*, Springer-Verlag, New York, 1999.

[25] R. A. Silverman, *Complex Analysis with Applications*, Prentice-Hall, 1974.

Index

Graduate Texts in Mathematics

(continued from page ii)

Printed In The United States Of America